Pioneers in Arts, Humanities, Science, Engineering, Practice

Volume 19

Series Editor

Hans Günter Brauch, Peace Research and European Security Studies (AFES-PRESS), Mosbach, Germany

http://www.afes-press-books.de/html/PAHSEP.htm
http://www.afes-press-books.de/html/PAHSEP_Picht.htm

Enno Rudolph · Johannes Picht
Editors

Georg Picht: A Pioneer in Philosophy, Politics and the Arts

Springer

INSTITUT FÜR
INTERDISZIPLINÄRE
FORSCHUNG
F·E·S·T Forschungsstätte der Evangelischen Studiengemeinschaft

UNIVERSITÄT
LUZERN

Editors
Enno Rudolph
Faculty of Cultural and Social Sciences
University of Lucerne
Lucerne, Switzerland

Johannes Picht
Schliengen, Baden-Württemberg, Germany

Acknowledgement:
Book cover photo provided with permission from the photo collection of the Picht family. More on Prof. Georg Picht, his books, other publications and links to selected media are found at: http://www.afes-press-books.de/html/PAHSEP_Picht.htm.

Translation grant from the Friedrich-Foundation (Hannover) and from the Hans and Gertrud Zender Foundation (Meersburg)

ISSN 2509-5579 ISSN 2509-5587 (electronic)
Pioneers in Arts, Humanities, Science, Engineering, Practice
ISBN 978-3-030-31792-8 ISBN 978-3-030-31790-4 (eBook)
https://doi.org/10.1007/978-3-030-31790-4

Copyediting: PD Dr. Hans Günter Brauch, AFES-PRESS e.V., Mosbach, Germany

This Springer imprint is published by the registered company Springer Nature Switzerland AG
The registered company address is: Gewerbestrasse 11, 6330 Cham, Switzerland

In grateful memory of the composer Hans Zender (1936-2019), friend and student of Georg Picht as well as patron of this edition

Acknowledgements

We would like to thank the Friedrich Foundation as well as the Hans and Gertrud Zender Foundation, without whose generous support this edition could not have been realized. We are equally grateful to Dr Christina Weiss, former Minister of State, whose open-minded advice opened the door to this possibility. We also thank the Protestant Institute for Interdisciplinary Research (FEST) in Heidelberg for making its infrastructure available for our work without any restrictions and for dealing with the administration of the budget.

Furthermore, we are grateful to the publisher Klett-Cotta and the Picht family for their cooperation in granting the rights to publish the texts. Finally, we wish to express our emphatic thanks to the translator, Wieland Hoban, both for rendering Picht's works in an English that does the greatest possible justice to his style and intentions, an achievement that can scarcely be overestimated, and for his great patience with our countless queries and emendations.

Lucerne, Switzerland — Enno Rudolph
Schliengen, Germany — Johannes Picht
March 2021

Introduction

When a renowned German weekly published a series of portraits of 'great German-language philosophers' of the twentieth century in 1971, a name that appeared alongside Karl Popper, Ernst Bloch, Max Horkheimer, Carl-Friedrich von Weizsäcker and others was that of Georg Picht. Only then was the general public – and the academic audience – suddenly made aware that a man they had hitherto perceived mostly as an expert on education policy and a participant in public debates on the politics of peace, the environment and science was in fact a philosopher.

Seven years earlier (1964), Picht had caused something of a stir with a series of articles entitled *Die deutsche Bildungskatastrophe* [The German Education Disaster], in which he accused politicians of failing in the field of education and called for wider access to academic education in all social strata. In other areas too, he had taken up politically sensitive topics and triggered controversies as well as intervening in ongoing disputes and pushing them in new directions. He was skilled in using the media – not only newspapers and magazines but also radio and television – to convey his political concerns to an increasingly large audience. But only a few had so far understood that these multi-faceted public activities stemmed directly from his understanding of philosophy. Before making philosophy his focus, he had first studied classics; he repeatedly referred to Plato as his most important philosophical teacher. Plato's insistence that philosophical insight should guide the life and organization of the polis became as central for Picht as the Socratic maxim that the truth of a philosophy must be proved not only by a person's words but also by their deeds. From Immanuel Kant, the Enlightenment thinker in whose work he kept trying to bring to light a coherent architecture, he adopted the demand that each individual must commit autonomously to using their own freedom in order to recognize the right of all people to freedom. Organizing the resulting reciprocal dynamics of actions in such a way that each individual person furthers the freedom of all by making use of their own freedom is the task of a community guided by reason. The use of freedom by a being endowed with reason is thus a public act, and aims for the creation of a critical public awareness. Only with this in mind can one understand the emphatic appeal to reason that repeatedly echoes in Picht's writings. He saw himself as a thinker in the Enlightenment tradition, albeit one who first of all had to confront the task of 'enlightening enlightenment about itself', that is, showing the preconditions – such as the assumption of achieving moral perfection among humans – underlying the programme of enlightenment. Thinkers such as Nietzsche and Heidegger, who relentlessly criticized Enlightenment philosophy, acted as counterpoints and models for Picht, and he engaged intensely with their work.

Kant had called for a 'revolution in the way of thinking', fundamentally challenging the validity of traditional metaphysics. As the originator of transcendental subjectivism and the author of both the *Metaphysical Foundations of Natural Science* and the *Metaphysics of Morals*, he had remained an avowed metaphysician on a new level - 'a metaphysician of metaphysics', as he put it. When Picht deals with Kant, he always bears in mind Kant's goal of bridging the gap between being and obligation, between theoretical and practical reason, and of formulating a philosophy that can bring to light the conditions of possibility of natural science and mathematics as well as those of the mature individual's political participation in the community. According to Picht, the fact that Kant encountered antinomies in this process that he was unable to resolve, yet did not evade them, demonstrates his greatness as a philosopher. One of these antinomies is what Kant considered an irreconcilable contradiction between the principle of consistent determinism of natural causality on the one hand and the principle of a 'causality out of freedom' (Kant) on the other. According to Kant, the former follows from the invariable validity of the 'a priori pure concept of the understanding', which he views as constitutive of the condition of possibility of experience as such, whereas the latter - in Kant's solution - is an idea of reason that both instructs the understanding to cognize the laws of nature by means of its pure concepts and is able to, and should, apply itself as a liberating cause of ethically responsible action.

In 1938, Picht was struck by the full force of the realization that the knowledge of nature enabled by the natural sciences gives humans the means not only to cognize the objective world of nature by formulating natural laws but also to destroy it using those same laws, when, through his friend Carl-Friedrich von Weizsäcker, he became one of the first outside observers to learn of the possibility of developing a nuclear bomb. Unlike his teacher Heidegger, however, he did not conclude from this that science in general and natural science, in particular, should therefore be rejected ('science does not think') but rather drew a twofold conclusion from the experience: first of all, it was necessary to define the area in which natural science (and thus *eo ipso* metaphysics, which lives on in its forms of thought and epistemological foundations) is true, in order then to ask in what function metaphysics continues to be effective. Picht thus demands an intraworldly interpretation of metaphysics, whose truth he does not negate, but whose claim to universality he reins in. Secondly, and no less philosophically, one had to ask after the conditions of possibility of human responsibility in history. Picht was convinced that this could not be restricted to writing scholarly essays about the matter for an expert audience; there needed to be a transition here from philosophical insight to political action.

It was consistent with this that as early as the 1950s, Picht was one of the first to raise his voice against the threat to the human *oikos* and its supporting political and social systems. When he was appointed as Professor of the Philosophy of Religion at the University of Heidelberg - the only such position in Germany to this day – in 1965, his characteristic combination of public activity as an academic and in the political and media landscape had been expressly welcomed as a criterion. When, as in a series of radio lectures broadcast in 1969 under the title *Mut zur Utopie* [Courage for Utopia] and subsequently translated into a variety of languages, he drew on detailed analyses and expertise to warn of the dangers of globalization and scarcity of natural resources, issues that were barely discussed at the time, he proved himself a cosmopolitan and public bearer of responsibility in Kant's sense, as well as a *zoon politikon* in the Platonic sense.

Kant and Plato are indeed the two philosophers who, more than any others, stand out for the way they attempted to bring together politics, science, ethics and mathematics in an overarching conception. In different (and complementary) ways, each of them places the human being in the centre and makes its self-awareness - which also means the awareness of its boundaries - their philosophical starting and vanishing point. In Kant, it is the self-cognition of the limits of reason which corresponds to a freedom that proves itself by only adopting maxims which condition their own possibility. Similarly, Plato understands freedom as the ability to impose rules of self-education upon oneself that enable an integration of the individual into a just state. Hence, both see freedom as autonomy, as self-legislation. This is not the only area in which Picht sees Kant as a successor to Plato, which exemplifies the way Picht repeatedly looks for the traces of antiquity in modernity and conversely reveals a form of enlightenment in classical philosophy that is no less significant than the modern Enlightenment. Both could be placed under Kant's motto: 'Enlightenment is the human being's emergence from its self-imposed immaturity'.

Another respect in which Picht sets himself apart from Heidegger, then, is that he leaves behind the latter's strictly-defended opposition of antiquity and modernity. He views the history of philosophy neither as the history of a progressive increase in knowledge that would permit the rejection of earlier philosophy as obsolete and inferior *a priori* nor as a series of opposing periods, let alone a history of degeneration. Rather, what characterizes his understanding of history is that every concept, every new form of thinking preserves the preceding forms and concepts within itself, representing and emerging from them - even where this is not based on a conscious reception of tradition. Consequently, if one is to understand philosophical concepts, they must always be made transparent against their historical background; for him, this understanding of what one is actually saying was the essence of philosophical education. What is required, then, is to reconstruct a figure of thought from the experience of thinking and the context from which it originally emerged; only by achieving this can one acquire the freedom to understand them in a new light, to examine them critically and develop their potential, rather than being at their mercy. That is why Picht was always concerned to gain insight into the present from the perspective of history but also to examine and rethink historical phenomena from a contemporary perspective. For him, systematic and historical thinking formed a unity. When he immersed himself profoundly in classical texts, it was not a matter of remote scholarliness but of recognizing the problems confronting the world today as clearly as possible from their origins. In this way too, the philosophical and the political were intertwined in his work.

One notable example of this is Picht's extensive interpretation of Parmenides, which is at once a meditation on the beginning of philosophizing. As one of the most important representatives of the pre-Socratic philosophers alongside Heraclitus, Democritus, Pythagoras and others, Parmenides had a decisive and lasting influence on Plato and Aristotle, and thus on the whole of Western philosophy. In the centre of his so-called 'didactic' poem stands the statement that Picht terms the 'equation of Parmenides' and acknowledges as the discovery of the principle of identity: *to gar auto noein esti te kai einai*, which he translates as: 'For the same Is: to think and to be' [*Denn dasselbe Ist: Erkennen und Sein*]. For Picht, this statement, with which Parmenides simultaneously proclaims the eternal presence of this being (staged as revelation by the goddess of truth) and sharply contrasts the knowledge thereof with any mere 'opinion' about transient matters, marks the birth of European metaphysics; he refers to the possibility of reading the works of Plato and Aristotle also as commentaries on Parmenides, thus following the reflections and alterations of the identity principle (as the fundamental law of metaphysics) through the history of philosophy and science up to the present day. Yet before that, he had described the development that leads from Homer via Hesiod and Xenophanes to Parmenides and becomes recognizable as an increasingly abstract reflection on the immortality of the gods and their omniscience, in contrast to the ever-limited knowledge of humans. From this origin, truth is tied to transcendence and to the principle of identity and marked as 'eternal' truth. While this concept of truth remains valid (e.g. as the 'eternal' truth of mathematical theorems), its exclusivity is no longer appropriate today. Other concepts of truth, likewise rooted in antiquity (e.g. in Heraklit), stand alongside it. Even the 'eternal' truth of Parmenides, then, must be interpreted within the horizon of a history of truth.

In such explorations, one witnesses and appreciates Picht's expertise as a philologist who not only had a mastery of the classical languages but also of the critical methods of compiling and securing texts as well as interpreting them, as a core skill that defines his work. Perhaps one could say that he treated the history of thought like a texture to be philologically processed, with different layers and threads of translation that had to be revealed and interpreted. Here, philology also has the important task of protecting the texts and their interpretation from 'corruption', that is, from being adapted to unsuitable needs and instrumentalized for ideological purposes. With this philological commitment to the integrity of texts from other periods, Picht also sets himself sharply apart – without engaging in polemic – from Heidegger, who often posited bold etymologies in his readings of classical philosophical texts in order to validate his own philosophy with their idiosyncratic results.

Picht not only dealt with philosophical texts in his investigations, however, but frequently also with works of literature, architecture and visual art, as well as music. He considered them equally valuable documents of thought and was convinced that art, especially in modernity, was capable of formulating insights that were often far ahead of academic philosophy, and that it reacted with greater seismographic sensitivity to tectonic shifts. He saw the most innate quality of art in its ability to offer an undisguised representation of the 'phenomenality of phenomena' – a concept whose importance in his late work is similar to that of the 'Being of beings' [*das Sein des Seienden*] in classical ontology. 'Art cannot lie', he writes in his lecture 'Art and Myth'.

Even today, forty years after his death, Picht is viewed as an exceptional figure in his discipline. As a researcher in the field of classical Greek philosophy, a critical analyst and intermediary for the legacy of the European Enlightenment as well as an educator and political adviser, he was one of the very few humanities scholars of the last few decades in the German-speaking world who was not only competent enough for a qualified, methodologically secure interdisciplinary dialogue, but also viewed this as the norm and the foundation of his own research and thought. As the long-standing director of the Protestant Institute for Interdisciplinary Research [*Forschungsstätte der Evangelischen Studiengemeinschaft*, FESt], he played a central part in developing up-to-date methods in this area. Philosophy as he understood it, namely as a mediator between historical competences across the boundaries of the scientific cultures treated separately today (natural sciences and humanities), and, on an even larger scale, as an attempt to bring together the spheres of science, art, religion as well as law and politics within a shared conceptual context, is more capable than any other discipline of organizing the diversity of specialist subjects, methods and perspectives found today inside a unifying matrix. It is the art of integration, which means not simply subsuming this diversity under the idea of a unity of knowledge, but rather to respond to the indissolubility of this diversity by taking seriously the incommensurability of the approaches whose structure it aims to reveal.

One can view Picht as a modern humanist of the old school who – like the Renaissance humanists from Petrarch to Erasmus – knew how modern antiquity was, and which antiquity was important in this regard. Yet he did not incorporate it in order to conserve it, but rather confronted its historical present with the present day in order to put it to the test. When selecting the texts for this volume, we tried to present a cross-section of his work and to document the most important areas of his thought as representatively as possible, though this meant that we had to pass over texts on particular issues (such as education policy). Picht's tribute to his friend Theodor W. Adorno acts as a prologue, and not only gives an impression of his perspective on Adorno's thought, but also demonstrates his philosophical interest in music.

For the first part, then, we chose three essays on classical Greek philosophy that simultaneously show Picht the philologist at work. The first has already been mentioned: in a close reading of the philosophical poem by Parmenides, it seeks to decode what was the first articulation of that original verbalization of the conception of truth which made its mark on European tradition for centuries, namely as the identity of thinking and being. The ontology of Parmenides also remained valid for Plato and Aristotle. Aristotle was faced with the task of mediating between the eternal presence of that which is and movement in nature; this is thematized in the third essay via an explanation of the concept of *energeia* that Aristotle introduced for this very purpose, alongside various other concepts. Placed between these two, Picht's study on the irony of Socrates is intended to offer at least a small sample of Picht's interpretations of Plato.

The second part comprises texts on ethics, politics and law. While the two essays 'Philosophy and Politics' and 'The Philosophical Concept of Ethics' take a fundamental and systematic approach to their respective subjects, albeit with a depth of historical insight, 'Kant's Transcendental Grounding of International Right' is an exposition of Kant's political theory as well as his theory of history. It stands alongside Picht's critical examination of the basis of legitimacy for the concept of universal human rights, to which - over 20 years after its initial publication – he added an afterword in response to the human rights policies of the Carter Administration.

One of the central themes of Picht's philosophical work, time, can only be documented in brief here, but it is present throughout the other texts. It forms the focus of two essays that we placed together in the third part of the book. The first - written for a festschrift on the 70th birthday of Werner Heisenberg - deviates from the metaphysical view of time in presenting time in terms of its phenomenality and showing the unity of time from the perspective of its differentiation in the three temporal modes, which Picht links to the three modalities of possibility, actuality and necessity. The second essay offers an approach to anthropology in terms of the historicity of the human being.

The texts in the fourth part are all taken from Picht's seminar series 'Kunst und Mythos' [Art and Myth] from 1972/73. They are intended to show the horizons of his thinking on art. 'Myth and Affect' belongs in this context, even though its presentation of classical mythologems would also have made it suitable for the first part of the book: for Picht, art and myth were equiprimordial phenomena, and he viewed it as superstitious to believe that one could emancipate oneself from myth.

The book closes with one of the last texts completed by Picht and published in his lifetime. It deals with 'the unthinkable as such', 'the power that opposes hope' and the forms in which evil, understood as such, hides in the very places where we think we have defeated it.

Lucerne, Switzerland — Enno Rudolph
Schliengen, Germany — Johannes Picht
March 2021

Contents

1 Prologue: Atonal Philosophy – In Memory of Theodor W. Adorno († 6 August 1969) 1

Part I Ancient Greek Philosophy

2 The Epiphany of the Eternal Present – Truth, Being and Appearance in Parmenides (1960) 7

3 The Irony of Socrates (1971) 43

4 The Concept of *Energeia* in Aristotle (1959) 57

Part II Ethics, Politics and Right

5 Kant's Transcendental Grounding of International Right (1971) 75

6 The Intellectual-Historical Background of the Doctrine of Human Right (1957/1980) 107

7 Philosophy and Politics (1972) 123

8 The Philosophical Concept of Ethics (1978) 129

Part III Time and History

9 Time and Modalities (1971) 151

10 The Historical Nature of the Human Being (1976) 161

Part IV Art and Myth (1972/1973)

11 The Conflict Between Art and Society 177

12 Absolute Art and Politics 189

13 The Crisis of European Art and the Emergence of Esthetics 197

14 Myth and Affect 203

15 Epilogue: On Evil (1981) 215

On Georg Picht 229

Selected Bibliography 233

On the Editors 235

Author Index 237

Subject Index 241

Chapter 1
Prologue: Atonal Philosophy – In Memory of Theodor W. Adorno († 6 August 1969)

If there were still such a thing as a history of spirit in Germany after Auschwitz, the death of Theodor W. Adorno would make it seem as if the clock had suddenly stopped.[1] In a society condemned to reproduce its own untruth, he broke the spell that shackled the freedom of the spirit. He called the delusional system that parades as reality by its true name, and could not be dissuaded from undertaking the 'powerless attempt of the idea to retain power over itself'. The recognition of this very powerlessness paved the way for a new autonomy of thought. Through its horror at the deception of affirmation, philosophy here gains legitimacy as a higher level of the critique in which, according to his interpretation, lies the truth of the history of European thought.[2]

Without any illusions about the inward and outward possibility of philosophy in the twentieth century, Adorno, resisting everything that is granted legitimacy, refused to evade the requirements asserted by the name 'philosophy'. Driven by fear, yet always fearless, the sensitivity of his thought was capable of achieving, every time it articulated itself, the impossible that he demands: 'It turns the quintessence of the experience accumulated in it to the objects, rends the veil with which society conceals them, and perceives them anew.' Over the last twenty years, Adorno changed the physiognomy of the intellectual landscape in Germany through the intensity of such spiritual experience. Anyone seeking to describe this period would note that its various tendencies can be precisely and sharply characterised by their stance towards that centre of innervation which Adorno's name represented, yet simultaneously concealed.

---

[1] This text was first published as: "Atonale Philosophie – Theodor W. Adorno zum Gedächtnis". In: *Merkur* 23 (1969), 889–892. Also in: G. Picht, *Hier und Jetzt: Philosophieren nach Auschwitz und Hiroshima*, Vol. I, Stuttgart: Klett-Cotta 1980, pp. 245–248. The permission to republish this text in English was granted by Klett-Cotta publishers.

[2] Theodor W. Adorno, 'Why Still Philosophy', in *Critical Models: Interventions and Catchwords*, trans. Henry W. Pickford (New York: Columbia University Press, 2012), p. 13.

E. Rudolph and J. Picht (eds.), *Georg Picht: A Pioneer in Philosophy, Politics and the Arts*, Pioneers in Arts, Humanities, Science, Engineering, Practice 19,
https://doi.org/10.1007/978-3-030-31790-4_1

The path of spirit is that of misunderstanding. Adorno did not find an adequate answer in the society he agitated. The vibration emanating from his thought was so strong, however, that even misunderstanding, even hostility and defamation, indeed open infamy, could not suppress it. The tremors he caused even proved capable – contrary to his own analysis – of affecting political consciousness and public opinion, for better or worse. That the claim 'the spirit is on the left' gained a political meaning for the first time in German history would be inexplicable without the effect of Adorno's thought.

The content and truth of that critical experience which he called 'negative dialectics' does not express itself primarily in the theoretical models that he used to raise sociology, musicology, literary studies, aesthetics and philosophy to the methodological level required by the objects of analysis. In the context of his thought, these models merely act as auxiliary constructions to enable an intellectual experience that cannot be captured by the schema of transferrable methods or 'positive' doctrines. His aim was not to enrich what we now call 'academic research' with new insights, methods and hypotheses within the established system of disciplines; this inevitable side effect contributed to a misrepresentation of his intentions. The primary content of his knowledge emerges whenever it transpires that this new form of critical thought cannot take place in a scholarship-immanent fashion, but rather seeks to question or even dismantle the entire framework of twentieth-century scholarship and its inherent premises.

One way in which this occurs is by showing that such scholarship, without admitting it to itself, reproduces the established social and economic conditions and their forms of domination merely through its 'pure' structure. It is a perversion of Adorno's thought, however, to remove this insight from the context of experience that contains its meaning and its truth, elevate it undialectically to a doctrine and, in an open betrayal of its author, force it into the diabolical claws of affirmation. Adorno's analysis was also confirmed by the fact that the very mechanisms he could unmask like no other even usurped his own insights. He bore it with the demeanour of a *grand-seigneur* who has no need for posturing because he has no reason to be ashamed of his sorrow.

If one wishes not only to acknowledge his work but to understand it, one must first realise that Adorno's dialectical thought, for reasons that cannot be expounded here, is so radically opposed to the positivity of logical structures that its dialectical movement even eludes the residuum of classical logic in Hegel, namely the schema of thesis, antithesis and synthesis. In the concept, it already seeks the contradiction that negates it as a concept and takes it apart. The consequence of this, which irrupts into the centre of thought and is therefore scandalous, is that philosophy here appears in a guise that not only eludes the choice between true and false upheld by the principle of non-contradiction, but in fact declares this choice itself 'false' in a different sense of the word. Anyone who sought to portray Adorno's philosophy as 'true' would already have denied this same philosophy. It identifies itself through its authenticity, which is expressed in the sensitivity with which the movement of micrological observation reacts to the matters it describes. It legitimises itself through the transparency of the ordeal contained in its experience. That is why it cannot be reproduced. Negativity, it transpires, is the path on which his thought opposes, step by step, the violence done to the truth by established science and philosophy through the very form in which they utter it.

Thus the structure of transcendental philosophy is transformed. Kant's discovery of transcendental semblance is no longer restricted to metaphysics; rather, it becomes clear that precisely the foundation of the truth of experiential knowledge defended to the last by Kant, namely the identity of transcendental subjectivity, rests on a transcendental semblance. The reflection of thought upon the one who thinks can only break through this transcendental semblance if it always also resolves the contradictions of subjectivity from which thought emerges, its needs and motivations, the constraints to which it is subject and the longing that is shackled by these constraints in the process of thought. The subject of this transcendental reflection is the empirical one, or, more precisely, the subject that is trapped in society's 'context of delusion' and must be freed from it. The discovery and liberation of this subject's subjectivity, which can only be carried out in the form of critique, is the content of spiritual experience that manifests itself in Adorno's work. His method is transcendental reflection to the power of three, reached via Marx (Kant representing the power of one and Hegel the power of two). As the truth of the subjectivity thus grasped can no longer rely on the unity of the transcendental subject and the schematicisms latent therein as a metaphysical foundation, transcendental reflection now requires all the instruments developed by the modern crisis sciences, especially sociology and psychoanalysis, for the critical illumination of empirical consciousness. But the intention here is always a liberation for possible experience; its path leads through the emancipation of society to the freedom of the spirit.

The horizon of this thought is determined by the content of the experience it enables, namely the horizon of subjectivity in a newly-discovered dimension of this concept that breaks through the coldness of bourgeois thought. Adorno distances himself as clearly from the metaphysics of absolute subjectivity as from all inferior forms of subjectivism. Yet even when it no longer makes itself explicit in the traditional metaphysical forms, subjectivity necessarily points to a kernel of light that appears in it and guarantees its truth. Experience only becomes transcendental if the *truth* of subjectivity can be demonstrated in verifiable experience. Adorno's thought assures itself of that truth through aesthetic experience, the experience of those trails and configurations, those antinomies, ruptures and eruptions in which the inner shape of artistic constructs manifests the traces of the life from which they have emerged. That is why music is at the centre of this philosophy; philosophy becomes indistinguishable from music. *Negative Dialectics* is an atonal philosophy.

Without explicitly stating it, Adorno explained the form of this work in the essay dedicated to Pierre Boulez, 'Form in New Music'.[3] Thought can assure itself of its truth through insight into artistic form because human beings cannot lie in art, even if they wish to lie; the sun brings it to light. A light shines out from the constellations of artistic form, a light from which the untruth is unable to hide. This is because form, as it emerges in all art, but most purely in music, should be understood as an

[3] Adorno, 'Form in der neuen Musik', in *Neue Rundschau* 77 (1966), pp. 19–34.

'intentionless language': 'The language of music is quite different from the language of intentionality. It contains a theological dimension. What it has to say is simultaneously revealed and concealed. Its Idea is the divine Name which has been given shape. It is demythologised prayer, rid of efficacious magic. It is the human attempt, however futile, to name the Name, not to communicate meanings.'[4]

Like music, Adorno's philosophy aims for precisely such an intentionless language. That is why his *Negative Dialectics* is the last book of European philosophy in which the great philosophical themes – the absolute, freedom and immortality, love and death – can unfold in all their hidden truth, protected by the form of negativity. Adorno himself formulated the cometary orbit of the spirit that is retraced by the experience of philosophy in such a way that its dazzling light renders even the hidden substrata of twentieth-century industrial societies transparent. The final sentence of *Negative Dialectics* reads: 'There is solidarity between such thinking and metaphysics at the time of its fall.'[5]

---

[4] Adorno, 'Music and Language: A Fragment', in *Quasi Una Fantasia: Essays on Modern Music*, trans. Rodney Livingstone (London and New York: Verso, 1998), p. 2 (translation modified).

[5] Adorno, *Negative Dialectics*, trans. E. B. Ashton (New York: Continuum, 1972), p. 408.

Part I
Ancient Greek Philosophy

Chapter 2
The Epiphany of the Eternal Present – Truth, Being and Appearance in Parmenides (1960)

In my essay 'Die Erfahrung der Geschichte' [The Experience of History] I developed the thesis that in the domain of Greek thought, truth is experienced as the epiphany of eternal presence.[1] I attempted to show how the form of truth manifested in this epiphany still determines, from the background and uncomprehended, the limits within which our experience of what actually is takes place. The epiphany of eternal presence turned into 'supratemporal' or 'timeless' truth, which developed its forms in classical logic and, through logic, also prefigured the structure and path of modern, objective and objectifying science. The path our thinking and our experience of the world had to take under the dominion of 'timeless' truth is the path of a negation of history that, in the Modern Age, led to an attempt to make history itself bow to logical reason through a total planning of legislation. This step from absolute truth to total planning is the final possibility, and simultaneously marks the close of the period that begins with the epiphany of eternal presence. We are therefore on the point of liberating ourselves from the negation of history and entering the human possibility of experiencing truth in history. Even if the experience of truth in history were an experience of the absolute in its appearance, it would still be determined by the negation in which the absolute constitutes itself as such. By contrast, the last part of my essay develops the proposition that time itself is being. If this proposition is potentially meaningful, then truth – as the truth of being – is the truth of time. Then truth itself has a history, and we experience the truth in history by experiencing the history of truth. Hence we cannot exit the historical period that stood in the light of the epiphany of eternal presence by taking the path of abandonment; this would in turn be a negation of history. Rather, we must first make it truly accessible for us as a

[1] This text was originally published as: 'Die Epiphanie der ewigen Gegenwart: Wahrheit, Sein und Erscheinung bei Parmenides'. In: H. Höfling (ed.), *Beiträge zu Philosophie und Wissenschaft: Wilhelm Szilasi zum 70. Geburtstag*. München: Franke 1960. Also in: G. Picht, *Wahrheit – Vernunft – Verantwortung: Philosophische Studien*. Stuttgart: Ernst Klett Verlag 1969, pp. 36–86. The permission to republish this text in English was granted by Klett-Cotta publishers.

E. Rudolph and J. Picht (eds.), *Georg Picht: A Pioneer in Philosophy, Politics and the Arts*, Pioneers in Arts, Humanities, Science, Engineering, Practice 19,
https://doi.org/10.1007/978-3-030-31790-4_2

period in our own history by understanding it as the period in the history of truth that enabled our own experience of truth. Accordingly, we ask after the original horizon of that manifestation of truth that sustains the ontology of Plato and Aristotle, and thus the whole of European science.

In the study 'The Experience of History', the assertion that truth appeared to the Greeks as the epiphany of eternal presence was reached via an interpretation of Aristotle's definition of the human being as a *zōon logon echon* – a creature that possesses the *logos*. The form of knowledge in which truth discloses itself as truth, namely *nous*, was 'consciously omitted' there because it did not belong in an investigation that sought to gain access to the experience of history by following the history of the concept of experience. The result of this omission was that the origin of the Aristotelian *logos* remained obscure, and that in the analysis of Kant, the shared ontological foundation of the concept of time and the concept of pure reason could be hinted at, but not revealed. The ontological foundations of Kant's thought are the ontological foundations of objective science. We cannot expect to attain the status of freedom in relation to this science and its effects on the shaping of the technological world if the foundation on which it rests remains hidden from us. That is why the question of the original sense of the 'epiphany of eternal presence' is one we must no longer avoid.

The following investigation does not actually seek to pose this question, only to prepare it. It will show that the epiphany of truth that determined the fate of European thought took place, in a sense that can be precisely defined, in the poem by Parmenides. In order to reach an understanding of Parmenides, the first part of the text will trace the path that leads from Homer via Hesiod to Xenophanes, whose concept of *nous* will be interpreted. The second part will attempt to build on this by examining some fragments from the poem. The method of our approach can only develop through a philological interpretation of difficult texts that have survived as fragments; there is no 'royal road' to the insights we seek. If we seriously mean to speak of the 'history of truth', we can no longer view the 'historical' as the 'relative' or the 'contingent', which indifferently circles a timeless truth. The well-worn distinction between 'historical' and 'systematic' work is then revealed as a naïve prejudice. We will only be able to explore and experience the history of truth if the great patience of philology is combined with the great patience of fundamental questioning.

For reasons of space, I have been forced to limit my consideration of existing research to the bare minimum. Experts will quickly notice how much I owe to the work of Hermann Fränkel in particular. Where I have taken a path of my own, I hope that the reason for this will be clear without a discussion of other views.[2]

---

[2] After the publication of this study, a considerable number of important contributions to the scholarship on Parmenides appeared and created a new situation. I believe the view presented here still stands in all important aspects, but it would have to be supported at greater length now. I therefore considered revising the text or adding notes, but it transpired that any such attempt would be thwarted by limitations of space. The study would have to be transformed into a book, and the 'systematic' aim of this essay, which proceeded from an examination of the foundations of the *Critique of Pure Reason*, would be obscured by the wealth of philological details. I therefore decided to republish

2.1 The Prehistory of Epiphany from Homer to Xenophanes

The main section of Plato's *Timaeus* begins with the following words (27 DE):

> *estin oun dē kat' emēn doxan prōton diaireteon tade: ti to on aei, genesin de ouk echon, kai to gignomenon men aei, on de oudepote? to men dē noēsei meta logou periplēpton, aei kata tauta on, to d'au doxē met'aisthēseōs alogou doxaston, gignomenon kai apollymenon, ontōs de oudepote on.*
>
> Therefore, in my opinion, it is necessary to begin by making the following distinction: what is that which always is that knows no becoming, and what is the ever-becoming that never is something that is?[3] The one can be encompassed through *noesis* with the help of the declaratory statement, as it is always being as per the same; the other, however, can be apprehended by an apprehension [*doxa*] through perception that cannot be revealed by a *logos* because it is something that comes into being and passes away, but never is in the manner of being.

The distinction observed here – the distinction between, on the one hand, that which always is, with the *noesis* that belongs to it, and on the other hand, that which is ever-becoming and ever-fading, to which the *doxa* belongs – not only forms the basis for the *Timaeus*. For Plato, it is the most fundamental of all distinctions; it is the original opening of the realm in which his philosophy unfolds.

It may therefore be surprising to find him saying that, in his 'opinion' [*doxa*], it is necessary to begin the investigation with this distinction; he thus seems to term it a mere *doxa*. All Plato means by this, however, is that this distinction must be considered a *doxa* until it is revealed – which occurs in a sense in the course of the investigation – in a *logos*. With the word *doxa*, then, he has something similar in mind to what Hegel wrote in the introduction to the *Phenomenology of Spirit*: 'But Science, just because it comes on the scene, is itself an appearance: in coming on the scene it is not yet Science in its developed and unfolded truth.'[4] Yet neither Plato nor Hegel manage to do away with the difficulty they both see by means of this explanation. For both of them find the foundation on which they are standing as a historical imprint that is given to them directly as 'appearance', in a form that allows no evasion in any direction. In their respective historical locations, they must recognize this appearance of knowledge and truth just as one must recognise what is evident a priori. But thought

---

the text with minor changes in the form in which it was written in 1959 as a birthday gift for my friend and teacher Wilhelm Szilasi.

[3] Editor's note: Parmenides consistently uses *estin* in the strictly ontological sense. He is not referring to facts in relation to possible predicates of things (*estin* as copula), then, but rather (aside from remarks about the misguided opinions of mortals living in semblance) exclusively to their being. That which is, is referred to by Parmenides as *to eon* (*ta eonta* in the plural). The translation must do justice to this consistent reduction of speaking about things to their mere being by restricting itself – even where this leads to unidiomatic formulations – to the verb 'to be' and its derived forms. As English (unlike Greek and German) lacks a nominalised participle, *to on* (*das Seiende* in German) is translated as 'that which is'. This maxim is also followed in quotations from Plato and Aristotle in so far as they refer to the Parmenidean ontological tradition.

[4] G. W. F. Hegel, *Phenomenology of Spirit*, trans. A. V. Miller (Oxford and New York: Oxford University Press, 1977), p. 48.

cannot deduce its own preconditions from itself a posterior. Therefore, no *logos* in the strict sense is possible in which this fundamental distinction has been revealed; for, conversely, this distinction is the condition of possibility for the *logos* as *logos*, and for every revealing act that seeks truth in a *logos*.

We must therefore seek the original meaning of this distinction adopted by Plato where it grew historically, namely in the thought of Parmenides. As we will see, it 'occurs' in the poem by Parmenides as epiphany, as underivable divine revelation; yet even this epiphany required a historical preparation. As we operate under the spell of supposed self-evidences today in our thinking and our view of historical phenomena, self-evidences that largely developed from misunderstandings of Greek ontology, it is indispensable for us as a preparation for entry into the thought of Parmenides to retrace, as far as possible, the historical path that leads to Parmenides.

We will begin with Homer's invocation of the Muses, whom Hesiod calls the daughters of Mnemosyne, of memory (*Theogony*, 53ff.). At the start of the Catalogue of Ships in Book II of the *Iliad* (484ff.), the poet interrupts the flow of the poem to call on the Muses for help:

Espete nyn moi, Mousai Olympia dōmat' echousai –
hymeis gar theai este, pareste te iste de panta,
hēmeis de kleos oion akouomen oude ti idmen –
hoi tines hēgemones Danaōn kai koiranoi ēsan.
Tell me now, you Muses who reside on Mount Olympus –
for you are goddesses and present everywhere, and have seen everything,
but we only hear the news and have seen nothing –
who were the leaders and rulers of the Danaans.

Here the divine knowledge of the Muses is contrasted with our human 'knowledge'. Only the goddesses can have true knowledge, for the perfect tense *oida* used by the Greeks to refer to knowledge means 'I have seen', 'I am an eyewitness'; the goddesses alone can be eyewitnesses to everything, because the goddesses alone are omnipresent. We humans, on the other hand, only hear news, we rely on hearsay, and lack the reliable knowledge to which an eyewitness can invoke. Thus in truth we hear, as stated in the first line of the *Iliad*, the voice of the goddess in the song of the poet. In Book II (488–492), Homer continues:

plēthyn d'ouk an egō mythēsomai oud'onomēnō ...
ei mē Olympiades Mousai, Dios aigiochoio
thygateres, mnēsaiath' hosoi hypo Ilion ēlthon.
But I could neither count nor name their multitudes myself [...]
if the Olympic Muses, the aegis-bearing daughters of Zeus,
would not call to my memory how many came before Ilion.

What does 'call to memory' mean here? It does not, at any rate, mean that the poet had already known the number and names of the heroes who came before Ilion but forgot them later, such that the Muses would simply be jogging his memory. Rather, the word *mnēsasthai* – 'call to memory' – is used because the relationship of the poet's indirect knowledge to the direct knowledge of the Muses corresponds to the relationship of the memory's indirect knowledge to direct apperception itself. In this sense, the Muses are the daughters of Mnemosyne: as conveyors of the immediate knowledge to which we humans only gain access through mediation.

How this mediation of the immediate is to be imagined is shown in the words of the singer Phemios in the *Odyssey* (22, 347ff.):

> *autodidaktos d'eimi, theos de moi en phresin oimas*
> *pantoias enephysen.*
> I am self-taught, but a god caused manifold paths to grow
> in my mind.

In a widely used metaphor, the 'manifold paths' are the paths of song; their meaning is shown in a line by Stesichorus (Fragment 25D): *ha Mousa gar ouk aporōs geuei to paron monon, all' epergetei panta therizomena,* 'for the Muse does not merely sample the present without a path, but reaches everything and collects the harvest.' To sample only the present does not require a path. But the Muse reaches everything, including what is absent from us humans – in the *Iliad* she is present everywhere and an eyewitness to everything – and thus requires manifold paths. The paths on which the Muse reaches 'everything' are the paths of song. And if the Muse 'plants' these paths in the poet's mind as if they had grown within him of their own accord (*enephysen*), he can report as if he had been present himself as an eyewitness and can offer up the harvest of the Muses as his own gift. This only applies, of course, to the self-taught singer, who does not sing the verses of other poets, but what the Muse herself has placed in his mind. The paths that open up in the poet's mind do so thanks to the instruction from the Muses known in Book II of the *Iliad* as 'memory'; but he finds them in his own mind and not someone else's, in his own *physis* (*enephysen*), and that is why, by human standards, he is self-taught. Thus we already find in the Homeric singer and poet the same agreement between divine memory, divinely revealed path, personal learning and personal natural disposition that would later return in Plato in the 'highest art of the Muses' (*megistē mousikē*, *Phaedo* 61 A), namely philosophy, as the doctrine of anamnesis, 'recollection' (re-remembering). In the doctrine of anamnesis, Plato discovers what would later be called a priori cognition. But the 'recollection' of a priori cognition is incomplete if it forgets that it originated from the inspiration of the Homeric poet by the Muses.

In Plato, the capacity for anamnesis is *nous*. We will therefore have to determine more precisely what is meant by *nous*, and how the *nous* of the later philosophy relates to the inspiration of the poet.[5] In Homer, the verb *noein* first of all means 'to espy', 'to see with the eyes' (*ophalmoisi noein*, in *Iliad* 15, 422). But because observing remains mere gaping if one does not also perceive what one sees with one's

---

[5] The following outline of the history of the philosophical concept of *nous* contradicts the study by K. Von Fritz, '*NOUS* and *NOEIN* in the Homeric Poems', in *Classical Philology* 38 (1943), pp. 79ff. There is insufficient space here for a comprehensive explanation; see now W. Luther, 'Wahrheit, Sicht und Erkenntnis in der griechischen Philosophie bis Demokrit', in *Archiv für Begriffsgeschichte* 10 (1966), pp. 13ff.

'inner eye', *noein* then means a perceiving apprehension. One example is sufficient to explain this usage: in Book XIX of the *Odyssey* (478f.), the wet nurse recognizes Odysseus by his scar. She looks for Penelope and wants to tell her that her beloved husband is in the house:

hē d'out' athrēsai dynat' antiē oude noēsai.
tē gar Athēnaiē noon etrapen ...
But she could neither look at him nor recognize (*noein*) him,
for Athena had turned away her inner eye (*noos*)

With this inner eye, one can even recognize things invisible to the physical eye, and one can even place distant things as vividly before one's eyes as if one had physically transferred oneself there. In Book XV of the *Iliad* (80ff.), this property of the *noos* serves as a comparison to show the wonderful speed with which Hera transfers herself from Mount Ida to the great Mount Olympus:

Hōs d' hot' an aixē noos aneros, hos t'epi pollēn
gaian elēlouthōs phresi peukalimēsi noēsē.
'enth' eiēn ē entha', menoiēēsi te polla,
hōs kraipnōs memauia dieptato potnia Hērē.
Like when the thoughts of a man who has been in many
lands take flight and his clever mind imagines, 'I want to be here, or there',
but he feels an urge towards many places:
thus flew Queen Hera with urgent haste through the air.

Of course, a human being, as indicated in these lines, can only see lands before their inner eye that they have already seen in person; if the Muses do not inspire them, they must rely on the ordinary form of memory, and their mind is subject to delusion (*atē*). Only the *noos* of Zeus is omnipresent and unclouded: *all' aiei te Dios kreissōn noos ēe per andrōn* – 'But always the *noos* of Zeus is stronger than that of humans (*Iliad* 16, 688). The *noos* of Zeus is 'the eye of Zeus', of which it is said:

panta idōn Dios ophthalmos kai panta noēsas
kai ny tad', ai k'ethelēs, epiderketai oude he lēthei.
the eye of Zeus, all-seeing and all-knowing,
gazes upon this too if he desires, and it does not remain concealed to him.[6]

Before the omnipresent eye of Zeus, from whom nothing remains concealed, everything shows itself in its *a-lētheia*.

In the observation of the *noos*, we encounter the same way of seeing as in the invocation of the Muses. Therefore, both can also be combined – by Hesiod, for example, in the passage from *Works and Days* (661/2) in which he seeks to explain the rules of seafaring. He himself, as he says, did not go to sea and can therefore not speak as an 'eyewitness'. He is in the same situation as the poet of the *Iliad* at the start of the Catalogue of Ships: *alla kai hōs ereō Zēnos noon aigiochoio; Mousai gar*

[6] Hesiod, *Works and Days*, 267f.

m'edidaxan athesphaton hymnon aeidein – 'but thus too I shall speak the *noos* of aegis-bearing Zeus;' (what the omnipresent *noos* of Zeus has before its eyes, namely the correct rules of seafaring) 'for the Muses have taught me to sing a song that breaks through my limitation.'[7]

A figure comparable to the poet is the seer, *hos ēde ta t'eonta, ta t'essomena pro t'eonta* – 'who (through the gift of Apollo) had before his eyes what is, what will be and what has been' (*Iliad* I, 70, of Calchas). We cannot be surprised, then, that this 'eternity formula', in the prooimion of Hesiod's *Theogony*, circumscribes the horizon to which the poet's spirit is moved by the inspiration of the Muses:

enepneusan de m'aoidēn
thespin, hina kleioimi ta t'essomena pro t'eonta,
kai m'ekelonth' hymnein makarōn genos aien eontōn.
but they breathed song into me,
divinely heralding song, so that I would praise what will be and what has been,
and told me to sing of the race of the blessed, of the ever-being.[8]

While these lines are based on the Homeric 'eternity formula', which can be followed through Hesiod himself (*Theogony*, 38) via Heraclites (B30), Empedocles (B21, 9) and Anaxogoras (B12) to Plato (*Parmenides*, 155d), they also adapt it in a highly significant way: the present is not placed together with the future and the previously existent, but is contrasted with them as the divine, which is eternal. Only the gods stand in the eternal presence of being; what comes into being and passes away in time, on the other hand, will never purely and completely attain presence, and thus beingness, on account of its temporality and ephemerality; it 'is' only ever in the transition from what previously was to what will be. But the gods themselves are in eternal presence, and all that *is* is present before their eyes. In Xenophanes, this unification of the two definitions then leads, in a major step, to a new perception of what we name using the word 'god'.

According to Xenophanes, 'from the start, everyone learnt from Homer' (*ex archēs kath' Homēron epei memathēkasi pantes*, B10) what the gods are. He himself presents the doctrine that everything Homer and Hesiod proclaim about the gods is based on the image of the deceptive and ephemeral nature of humans, and hence a vain delusion. He could have invoked Hesiod himself for this purpose, for the Muses, in their epiphany, told him:

idmen pseudea polla legein etymoisin homoia,
idmen d', euth' ethelōmen, alēthea gērysasthai.
We know many deceptive things to say that look true,
but we also know, if we want, how to report the truth.[9]

---

[7] Regarding *athesphaton* ('unlimited, boundless'), see Hermann Fränkel in *Antidoron, Festschrift für Wacknernagel* (Göttingen: Wandenhoeck, 1923), pp. 281f.

[8] Hesiod, *Theogony*, 31f.

[9] Ibid., 27f.

This means that even song inspired by the Muses can contain some deception. In Hesiod, because the Muses who appeared to him at the foot of Mount Helicon gave him the certainty that he would tell the truth about the gods, these verses stand in contrast to other poets, whose reports of the gods only appear to be genuine, though they too speak with the voice of the Muses. But according to Xenophanes, Hesiod was also deceived, because, like Homer, he blamed the gods for the blasphemy of humans (B11), and because he teaches that the gods are created, even though they are *aien eontes*, ever-existent (B14). Inspiration by the Muses is no guarantee of truth. He therefore says (B34):

kai to men oun saphes outis anēr iden oude tis estai
eidos amphi theōn te kai hassa legō peri pantōn;
ei gar kai ta malista tychoi tetelesmenon eipōn,
autos homōs ouk oide; dokos d'epi pasi tetyktai.

No man has ever seen what is immediately clear, nor will there ever be one who [as an eyewitness]
would have knowledge about the gods and of what I say about all things;
for even if it occurred to the fullest extent that someone spoke in a concluding fashion,
he would not know it from his own observation, for a semblance is created over everything.

We will see later that this semblance (*dokos*) stands in contrast to the truth of the divine *nous*; this means that this fragment by Xenophanes contains the first instance of the distinction between *nous* and *doxa* that we encountered as the fundamental ontological distinction in Platonic philosophy. We must therefore examine this fragment more closely. The theme of the poem that contains this fragment becomes apparent in the second line: it is about the gods and 'everything'. 'Everything', *panta*: this is a constantly recurring description of the entire realm of *physis*. Hence these lines are rightfully given the heading 'Peri physeōs' – on *physis* – among the fragments of the pre-Socratics, and there should be no doubt that Xenophanes was the first to write a poem about this topic, and thus provided the model for his student Parmenides. Admittedly, he follows on directly from Hesiod, for in the lines cited above (*Theogony*, 31f.), in which Hesiod names the topic of his own poem, the words *ta t'essomena pro t'eonta* ('what will be and what has been') correspond to the words *peri pantōn* ('about everything'), and the words *makarōn genos aien eontōn* ('the race of the blessed, of the ever-being') to the words *amphi theōn* ('about the gods'). The connection to Hesiod becomes even clearer through a further fragment that surely belongs in the same context (B35): *tauta dedoxasthō men eoikota tois etymoisin* ('this shall be presented as a *doxa*; it resembles the genuine'), for this line contains a literal allusion to line 27 of the *Theogony* quoted above: *idmen pseudea polla legein etymoisin homoia* ('We know many deceptive things to say that look true'). This twofold connection to Hesiod's prooimion on the Muses confirms that we were on the right path when we proceeded from the poet's invocation of the Muses in order to establish the distinction between *nous* and *doxa*.

Yet it is precisely on recognizing the allusion to Hesiod that it becomes clear how the same words mean something entirely different from the mouth of Xenophanes. Certainly the poet who, with such unprecedented self-confidence, showed that Homer and Hesiod, the teachers of the whole of Greece, were guilty of unworthy deception, did not mean to show with these words that the Muses had deceived him as well. Rather, he wanted to say that even a poet can achieve nothing greater with his words than a semblance that seems identical to the real thing. In order to demonstrate this, he takes the motif from the invocation of the Muses in Book II of the *Iliad*: the poet himself is not an eyewitness, and therefore possesses no authentic knowledge, especially not of things that exceed any human measure, such as the gods or the cosmos. From the invocation of the Muses, however, Xenophanes only adopts the insight into the limitations of human knowledge; the Muses who could provide help have disappeared – they deceived Homer and Hesiod, after all. Where the Muses should appear, we read the meaningful words: *ei gar kai ta malista tychoi tetelesmenon eipōn, autos homōs ouk oiden* – 'for even if it occurred to the fullest extent that someone spoke in a concluding fashion, he would not know it from his own observation'. *tetelesmenon* – 'something that has been brought to completion': this means that what is said is self-contained and has achieved its goal. The goal that allows a speech to conclude lies not in the complete roundedness of the speech itself, however, nor in the inner cogency of the thoughts. When Homer states in a speech, *to de kai tetelesmenon estai* (*Iliad* I, 212), this means that the speaker's promise is kept, that his plan is indeed put into practice. Accordingly, Xenophanes's comment means that the speech has achieved its goal – the truth – and 'fulfils' itself in this sense. So we find the marked paradox that even true speech is still mere semblance, for the speaker cannot know what they say truly, as an immediate eyewitness. One can only say, 'it happens that he speaks the truth'. Thus truth becomes a gift from Tyche. In Xenophanes, the words *ei tychoi* – 'if it occurred' – occupy the place where the Muses stood in Homer and Hesiod. Even in Plato, one often finds a 'divine Tyche' (*theia tychē*) in the same place.

If this is the case, then one can indeed only conclude that *dokos d'epi pasi tetyktai* – 'a semblance is created over everything'. Linguistically, this is a bold construction; one can see that the poet wishes to provide a compressed formula. The word *dokos*, which appears only twice more in the surviving literature by late writers probably dependent on Xenophanes, seems to be a neologism coined by Xenophanes for this passage. This points to the fact that Xenophanes had something new in mind that could not be described in familiar words. If the words *epi pasin* – 'over everything' – meant what Xenophanes goes on to present, namely the words of the poet, a neologism would not have been necessary, as *doxa* was available for this purpose. It is a different matter if *epi pasin* means the same as *peri panton* two lines earlier, namely everything that comes into being and passes away in *physis*, Hesiod's 'what will be and what has been' (*ta t'essomena pro t'eonta*). Then *dokos* is not the 'subjective' opinion of humans, but rather belongs to the essence of *physis* itself. This interpretation is confirmed by the word *tetyktai* – 'is created'. *teuchō* means 'to fashion artfully'; if *dokos* were the poet's 'illusory opinion' (as Diels argues), these words would mean that the poet had used his art to create a deliberate illusion. If one translates *dokos* as 'acceptability' [*Annehmbarkeit*] (like Fränkel), the contrast with truth is lost. But *teuchō* is often used in Homer and Hesiod to describe the artifice of the gods when they cause something – not only when, as with Hephaestus, the matter is the creation of a work of art, but also in natural processes caused by the gods: Zeus 'creates' rain, hail and snow (*Iliad* X, 6). Hence the word can also refer to something that 'happens', without anyone who could be called the author of the event: the opened wings of a door 'create' light (*Iliad* XXI, 538). Thus the *dokos* is 'created' from the essence of *physis* over everything that comes into being and passes away in *physis*, as a semblance that emerges from *physis* itself. This interpretation of the words is confirmed by the fact that it builds the bridge to Parmenides.

Against the background of this profound thought, the words in Fragment B35 take on a new and unexpected meaning. *tauta dedoxasthō men eoikota tois etymoisin* – 'this shall be presented as a *doxa*; it resembles the genuine'. We saw that Xenophanes alluded here to the line from Hesiod, *Theogony*, 27: *idmen pseudea polla legein etymoisin homoia*. In Hesiod, this line is aimed at other poets and their deceptive accounts of the gods, which he opposes with his own true account. As Xenophanes now also accuses Hesiod of deception, one would expect that he would counter the supposed truth of Hesiod with his own higher, pure truth. What does it mean that he instead adopts the very words Hesiod aims at the deception of poets, and applies them to his own poem? The only possible answer is a paradox: the semblance as such, that is, the discovery of the semblance – this is the higher truth of Xenophanes. Hesiod fell prey to deception precisely because he claimed to be reporting the truth. Xenophanes, on the other hand, claims to be reporting the semblance – *tauta dedoxasthō* – and thus becomes able to speak the truth. For if the *dokos* is created over everything that comes into being and passes away in *physis* itself, then the understanding and articulation of this *dokos*, namely *doxazein*, is the form of speech that corresponds to *physis*. Then *physis* is a domain in which the truth appears, but only appears, a domain in which it can never show itself as it is in itself, in its immediate clarity (*saphes* B34, 1) and authenticity (*etyma* B35). And so the realization that we are at the mercy of the *dokos* proves the only form in which we can recognize the truth. The paradoxical way in which Xemophanes points to the truth of the semblance that he has discovered is profoundly connected to the paradox in which Socrates later articulated his knowledge: *synoida emautō ouden epistameno* – 'I appear myself in the trial of Elenxis against me as cognizant of the fact that I am one who knows nothing.'[10]

In our elucidation of this fragment, we have so far omitted what may be the most important word. It is the unassuming *men*. The *men* places the *doxazein* of the poet on one side of an opposition; only the divine *nous* could have been on the other side. The reference to the truth lies in the *dokos* itself, for anyone who recognizes it for what it is, namely as mere semblance, must recognize the link with the truth, for which Xenophanes uses the word *eoikota*: the opposition in agreement and agreement in opposition that we describe with the unclear term 'similarity', and with which the

[10] See Georg Picht, 'Wissen des Nichtwissens und Anamnesis: Der Übergang von Sokrates zu Platon', in *Wahrheit, Vernunft, Verantwortung* (Stuttgart: Klett-Cotta, 1991), pp. 87ff.

Greeks defined the relationship of semblance to truth. Hence Xenophanes, through his insight into the nature of the *dokos*, is entitled to speak of the divine *nous* even though no human, as he says, has knowledge of the divine *nous* based on their own observation (B34). We possess a small number of fragments that show us an outline of his theory.

We begin with a fragment (B24) which is meant to show that there is good reason to speak of a divine *nous* in Xenophanes: *oulos hora, oulos de noei, oulos de t'akouei,* 'He sees as a whole, he thinks as a whole, he hears as a whole'. When he sees as a whole, he is wholly eye; when he hears as a whole, he is wholly ear; and if he is both at once, and wholly so, this should not be taken literally, but rather points to the meaning of the word in the middle: *oulos noei* – 'he is wholly *nous*'. As a whole: this means that he is pure *nous* through and through, and nothing else. The divine *nous*: according to the line by Hesiod cited above (*Theogony*, 267), *panta idōn Dios ophthalmos kai panta noēsas* – 'the all-seeing and all-knowing eye of Zeus', from which nothing remains hidden (*oude he lēthei*, in *Theogony*, 268), but which sees everything directly as it is. In order to see everything directly as it is, the *nous* of Zeus, like his mediators, the Muses (*Works and Days*, 661/2), must be present everywhere (*Iliad* II, 485). The *nous* of humans already has the possibility of travelling to a great variety of areas in an instant, and thus being present in several places at once. But this is only the indirect present of a recalling-to-oneself of what one already knows (*Iliad* 15, 80ff.). The eye of Zeus, on the other hand, sees everything and sees it simultaneously and directly; hence the nature of the *nous* is direct omnipresence. But if he is omnipresent as a whole (*oulos*, B24) then he cannot move himself. Fragment B26 tells us:

> *aiei d'en tautō mimnei kinoumenos ouden,*
> *oute meterchesthai min epiprepei allote allē.*
> He always abides in the same, not moving at all,
> and it does not befit him to go now here, now there.

The second line of this fragment is directed against the anthropomorphic form of the Homeric gods, who move now here and now there in an instant, and, if they are far away in the land of the Aethiopians, will not notice what is happening on earth (*Odyssey* V, 282ff.). But where is the divine *nous* located if, unlike the human *nous*, it is not fitting (*Iliad* XV, 80ff.) for him to move about? Xenophanes says: *aiei d'en tautō mimnei* – 'he always abides in the same'. Diels translates this as 'in the same place' [*am selben Ort*], but this translation is nonsensical, for the 'place' of the *nous* is omnipresence; it is everywhere at once as a whole. The true meaning of the words becomes clear from a passage in Parmenides in which, as Diels in particular saw, he interprets Xenophanes's 'original' (B8, 29f.):

> *tauton d'en tautō mimnei kath'heauto te keitai,*
> *choutōs empedon authi menei.*
> As the same within the same, it abides and rests in itself,
> and thus it will abide there immovably.

Hence the words of Xenophanes mean that the *nous* remains in itself, in the omnipresence that constitutes its essence. In contrast to the truth of the omnipresent *nous*, the variety of places and areas belongs in the realm of the *dokos*, the semblance, created over everything. The realm of semblance also includes the variety of times, however, for we have already seen from the 'eternity formula' since Homer that the knowledge conveyed by the gods and Muses, the knowledge of the divine *nous*, equally encompasses *ta t'eonta, ta t'essomena pro t'eonta* – what is, what will be and what has been. 'Knowledge', as we have seen means to have something present before one's eyes. For the knowledge of the *nous*, the past and the future are also present. Thus the *nous* is in all places and all times at once, and as a whole. In every direction, it is eternal present. This interpretation too is confirmed by Parmenides. It is now clear that the *nous* cannot move. For it neither has the possibility of moving from place to place (*phora*) nor can it, which is always and everywhere entirely itself, change (*alloiōsis*), for change is the transition from an earlier to a later state – but for him, the 'earlier' or the 'later' is still the One eternal present.

2.2 The Epiphany of Truth in Parmenides

We explained the words by Xenophanes *aiei d'en tautō mimnei* – 'he always abides in the same' – with the help of a passage by Parmenides of which, citing Diels, we claimed that it is based on Xenophanes. But was this really justified? Does not Xenophanes speak of the divine *nous*, while Parmenides means being? And is there not agreement in recent research that, precisely because of this difference the 'philosopher' Parmenides should not be viewed as a student of the 'poet' Xenophanes? We could counter this by referring to ancient tradition, for the testimony of Plato (*Sophist*, 242c-d) and Aristotle (*Metaphysics*, 986b22) is not refuted by the arguments adduced against it. But it would be even better to find the answer in Parmenides's own words, for which Xenophanes was meant to prepare us.

In Parmenides we find the following lines shortly after the passage just cited (B8, 34ff.):

> *tauton d'esti noein te kai houneken esti noēma.*
> *ou gar aneu tou eontos, en hō pephatismenon estin,*
> *heurēseis to noein; oud' ēn gar ē estin ē estai*
> *allo parex tou eontos, epei to ge Moir' epedēsen*
> *oulon akinēton t'emenai; tō pant' onom' estai,*
> *hossa brotoi katethento pepoithotes einai alēthē,*
> *gignesthai te kai ollysthai, einai te kai ouchi,*
> *kai topon allassein dia te chroa phanon ameibein.*
> But the same is: to think, and of what is the thought.
> For you will not find thinking without that which is,
> in which it is uttered; and there was not, is not and never shall be

anything besides what is, since fate has bound it
so as to be whole and immovable; therefore all things
will be mere names that mortals have defined, believing them to be true:
becoming and passing away, being and non-being,
change of place and alteration of bright colour.

The theme of these lines is named with the first word: *tauton* – the same. The same that is meant here must be the same of what it was said a few lines earlier (l. 29): *tauton d'en tautō mimnei* – 'he always abides in the same' – that same, then, in which, if we are right, the *nous* of Xenophanes abides. In Parmenides, however, unlike Xenophanes, this same is unfolded; it appears as the same in two different parts, as identity in difference. That is why it appears twice in line 29: *tauton* (the same) and *en tautō* (in the same). Accordingly, it appears in line 34 in two parts joined by *te kai*: *noein* (thinking) and *houneken esti noēma* (of what is the thought). Then, in addition, this sameness is explained by rejecting the separation of the two parts: *ou gar aneu tou eontos [...] heurēseis to noein* – 'for not without that which is [...] will you find thinking.' So the second part, which contrasts with thinking (*noein*) and is 'that of what is the thought', is 'that which is', *to eon*. For Parmenides, as we will see later, 'that which is', *to eon*, is not something that is next to another thing that is, but the whole of the 'well-rounded' truth of being. In the following, when we speak of 'that which is', we will place the word in quotation marks. Line 34 therefore says the same thing as Fragment B3 in elaborated form: *To gar auto noein estin te kai einai* – 'For the same Is: to think and to be.'[11]

These are initially only formal observations, and do not yet explain what these lines mean. But if we cling to the connection with Xenophanes, we arrive at an almost startlingly simple assumption. Xenophanes had said of God: *oulos noei* – 'he thinks as a whole'; he consists of nothing but this thinking. Parmenides says: *to gar auto noein estin te kai einai* – 'For the same Is: to think and to be.' Is it far-fetched to suppose that thinking (*noein*) in Parmenides is the same thinking of which Xenophanes speaks, namely the thinkingof the divine *nous*? Then Parmenides would have taught that the thinking of the divine *nous* of which Xenophanes speaks is the same as being.

This assumption, if correct, has far-reaching consequences for our understanding of Parmenides. We must therefore examine the connections between Parmenides and Xenophanes even more closely using the lines cited above. The group of lines has a strictly symmetrical structure: the first three and a half lines (from 34 *tauton* to 37 *eontos*) are, as we saw, an interpretation of *tauton*, 'the same', in which the simultaneously similar and different parts, namely *noein* ('thought, thinking') and *eon* (that which 'is'), abide, as the preceding group (26–33) taught. Exactly in the middle, these are followed by two half-lines (37–38) that, with reference to the preceding group, explain the state of abiding in the same: *epei to ge Moir' epedēsen oulon akinēton t'emenai* – 'since fate has bound it so as to be whole and immovable'. Fate,

[11] Cf. Laks/Most (ed./transl.): Early Greek Philosophy V/2, Loeb Classical Library vol. 528, Harvard University Press 2016, p. 39: 'For it is the same, to think (*noein*) and also to be' (ed.).

Moira, is the 'mighty necessity' (*kraterē Anankē*) that through its bonds, according to line 30/1, keeps the abiding 'in the shackles of its boundary', that is, in its abiding. Finally there are another three and a half lines that dismiss the 'opinions of mortals' (*brotōn doxai*, B1, 30): becoming and passing away, change of place and 'alteration of bright colour' are mere 'names', mere 'words' that are not 'uttered in that which 'is'' (l. 35). The last group of lines will require an explanation. It is already clear, however, that lines 34–41 convey the entire structure of Parmenidean thought, for they lead from the same in thinking and in that which 'is', via the bonds of Moira, to the realm of *doxa*. This makes it all the more significant for our question as to the relationship of Parmenides to Xenophanes that the structure of these lines follows on directly from Fragment B26 by Xenophanes in three ways:

1. Lines 34–37 *eontos* correspond to the words *aiei d'en tautō mimnei*
2. The two half-lines 37/38 – *epei to ge Moir' epedēsen oulon akinēton t'emenai* – correspond to the words *kinoumenos ouden* (the term *oulon* comes from Xenophanes B24)
3. The last group of lines corresponds to the words *oude meterchesthai min epiprepei allote allē*, for in *oude [...] epiprepei*, Xenophanes rejects the opinions of mortals, and *meterchesthai* is taken up by *topon allassein* – 'change place'. If one places the Xenophanes next to them, the lines by Parmenides read like an exegetical paraphrase – which is not to question the fact that Parmenides's thought operates in an area never reached by Xenophanes. At any rate, we are now justified in viewing the relationship of Parmenides to Xenophanes as certain.

We have taken the path that led us to Parmenides in order to uncover the ontological foundation of the Platonic distinction between *nous* and *doxa*, between *ousia* and *genesis*. We will therefore only go as far in our interpretation of Parmenides as is necessary for this particular issue.

At the end of the prooimion (B1, 28ff.), the goddess speaks to the poet:

chreō de se panta pythesthai,
ēmen alētheiēs eukykleos atremes ētor
ēde brotōn doxas, tai souk eni pistis alēthēs.
It is necessary for you to learn everything,
both the unshakable heart of well-rounded truth
and the opinions of mortals, in which there is no true reliability.

These lines appear in a context that is familiar from the Homeric invocation of the Muses, the call to the poet in the prooimion from Hesiod's *Theogony* and Fragment B34 by Xenophanes. The Homeric invocation of the Muses at the start of the Catalogue of Ships provides the insight that the opinions of mortals are not truly reliable: *hēmeis de kleos oion akouomen oude ti idmen* – 'but we only hear the news and have seen nothing (know nothing)'. In Homer it was the Muses, omniscient through their omnipresence, who conveyed true knowledge to the poet. In Hesiod too, the Muses reported the truth, but he knows that they are also capable of saying deceptive things that only appear true. In Xenophanes the Muses disappear, because he realizes that all conveyed knowledge is deception, even if it corresponds to the truth, because humans never see the truth as eyewitnesses; their only access to the truth is to recognize the illusory nature of semblance, for only semblance is directly given to them. The Muses do not come to Parmenides either, as they came to Hesiod at the foot of Mount Helicon; rather, he himself is carried off in his mare-drawn carriage. Unlike Xenophanes, he is spoken to again by the goddess, but this goddess – unlike the Muses – is not a mediator but *Alētheia*: she herself is immediate truth. This also results in a new poetic form. Homer begins the *Iliad* with the words *mēnin aeide, thea* – 'sing, goddess, the rage [...]'. Hesiod tells us (*Theogony*, 31f.) how this is to be understood: *enepneusan de m'aoidēn /thespin, hina kleioimi* – 'but they [the Muses] breathed song into me, divinely heralding song, so that I would praise [...]'. The song is breathed by the Muses, yet it is the poet himself who sings. Even if a god lets the paths of song grow in him, he is *autodidaktos*, 'self-taught' (*Odyssey* XXII, 437f.). Xenophanes speaks emphatically as himself; he confesses the deception of his human knowledge (B34, 1/2). In Parmenides, however, *Alētheia* herself speaks: she cannot make use of a mortal voice.[12] We therefore receive the lesson of this poem in the new form of an immediate address from the goddess to the poet. We are not simply told at the start, as in Hesiod, that the Muses appeared to the poet; rather, the poem itself is a direct epiphany. In the face of this poetic form, it is almost blasphemy to ask why Parmenides 'made use' of the form of a didactic poem for his 'abstract' doctrine. For if the epiphany that other poets can only recount appears here in a form that lets it occur directly, this poetic form in Parmenides is not only a 'medium', that is, mediation, as for the other poets, but rather the immediate matter itself. Epiphany and nothing else *is* the lesson, and anyone who explains it as a mere 'poetic' form thus also transforms the lesson into a mere *doxa*.

As it is truth itself that speaks, the semblance and the deception of humans are made manifest not by humans, but by the truth. What this means is also evident in a comparison with the words of the Muses to Hesiod (*Theogony*, 27f.):

> We know many deceptive things to say that appear genuine,
> but we also know, if we want, how to report the truth.

[12] It is commonly thought that the goddess Alētheia was introduced by Pindar (see *Olympian* X, 3f.; Martin P. Nilsson, *Geschichte der griechischen Religion*, vol. 1 [Munich: C. H. Beck, 1941], p. 706). But Parmenides names her in B2, 4 and B1, 29. It is clear from the following interpretation that she is the goddess who takes up the poet in B1, 22 and speaks everything that follows. Here we will only note what can be deduced from B1: 1. She is identical to the goddess in B3; the 'much-sung way of the goddess' (2/3) is the 'way of Alētheia' of which Pindar speaks (*Pythian* III, 103 and I. II, 10) . 2. She sends the daughters of the sun, who take the poet out of the house of night and into the light, pushing the veils from their heads as they leave the house of night (10). This is a mythical image representing the crossing over from the concealment of night into the realm of Alētheia. 3. While she is invoked together with the Muse in Pindar (*Olympian* X, 4), she here replaces the Muse. The prior history of the motif reveals that only Alētheia can replace the daughters of Mnemosyne. 4. The identification with Dike is impossible, for Dike opens the gate through which the carriage passes, only reaching the goddess in line 22. – The religio-historical origin of the goddess Alētheia should be sought in the domain of mystery religion, from which Parmenides adopts the pictorial language of the prooimion; there is insufficient space here to explain this assumption.

In Hesiod, as we saw, the first line referred to the deceptive teachings of other poets and the second to his own true teaching. We also saw how Xenophanes, for whom any utterance of the Muses is automatically deception, referred in paradoxical fashion to his own teaching (B35): *tauta dedoxastho men eoikota tois etymoisin* – 'This shall be presented as a *doxa*; it resembles the genuine'; and we saw how this appropriation of the *doxa* takes on an unexpected meaning through the understanding of the *dokos* (B34) that is created over everything. The *doxazein* understood in its essence was the form of knowledge assigned to the recognized *dokos*; in this sense, it was the true form of knowledge. In Parmenides, this new perspective also places the two lines by Hesiod in a new light. 'It is necessary for you to learn everything', says the goddess:

both the unshakable heart of well-rounded truth
and the opinions of mortals, in which there is no true reliability.

The first of these two lines corresponds to the second line of Hesiod, which had to be omitted by Xenophanes but is now restored to its original significance because *Alētheia* herself speaks it. The second line, as we saw, does not initially refer to Hesiod but to the second line from the corresponding couplet in the Homeric invocation of the Muses (*Iliad* II, 686). And yet there is a complete shift that only became possible via Xenophanes: here we do not find a human lamenting the deceptive nature of human knowledge; rather, the opinions (*doxai*) of mortals are revealed as necessary by the goddess, that is, by truth itself. So we do not encounter mere semblance, as in Xenophanes, but rather an insight into the nature of semblance. This is the sense in which Xenophanes had related the first line of Hesiod to himself. It is also the sense in which the line by Parmenides about the *doxai* of mortals shows a surprising connection to Hesiod. While the Muses told Hesiod, 'We know many deceptive things to say that appear genuine', what is now revealed to Parmenides in the opinions of mortals is the truth of semblance, whose nature is that it is not genuine, but appears genuine. Hesiod only related the second line of the Muses' address to himself: the proclamation of the truth. Xenophanes only related the first line to himself: the recognition of semblance. Parmenides can adopt both lines, for both things are revealed to him: that which is true and that which looks like the authentic. That is why the goddess says to him: *chreō de se panta pythesthai* – 'it is necessary for you to learn *everything*'. It is necessary: these are no empty words, but must equally be understood from the perspective of the truth. The necessity mentioned here (which in Parmenides is connected to Dike, Ananke and Moira) arises from the nature of truth itself; he must learn everything because it is impossible to learn truth itself without the truth of semblance, or to recognize semblance as semblance without truth itself.

We spoke just now of truth itself. Parmenides refers to it instead as *Alētheies eukykleos atremes ētor* – 'the unshakable heart of well-rounded (divine) truth'. What do these words mean? We will seek an answer by examining the meaning of the words in which the nature of *Alētheia* is elaborated here.

Truth is referred to as *eukyklēs*, 'well-rounded'. Parmenides explains this in the lines that close the first part of the poem, the revealing of truth (B8, 42ff.).

autar epei peiras pymaton, tetelesmenon esti
pantothen, eukyklou sphairēs enalinkion onkō,
messothen isopales pantē; to gar oute ti meizon
oute ti baioteron pelenai chreon esti tē ē tē.
oute gar ouk eon esti, to ken pauoi min hikneisthai
eis homon, out' eon esti hopōs eiē ken eontos
tē mallon tē d'hēsson, epei pan estin asylon.
hoi gar pantothen ison homōs en peirasi kyrei.

As merely translating these lines will not help us, we will combine a piecemeal translation with a brief explanation. L. 42/3: *autar epei peiras pymaton, tetelesmenon esti pantothen* – 'But as there is an outermost limit, it [that which 'is'] is brought to completion from every direction'. Parmenides had already said in line 30/1 that mighty necessity (Ananke) keeps that which 'is' in the shackles of the boundary that encloses it on all sides; there he continues: *houneken ouk ateleutēton to eon themis einai* – 'because of this, it is not permitted for that which 'is' to be incomplete [*ateleutēton*].' 'In ancient language, the word *ateleutēton* never refers to spatial limits, but always means: not completed, not realized, not brought to its goal.'[13] The boundary that necessity imposes on the 'existent' is thus not a spatial boundary, but rather the necessity to be always fully realized, always fully present or – in Aristotle's interpretation – always pure *energeia*. It is a mighty necessity that keeps that which 'is' within the shackles of its boundary in this sense, for only thus that which 'is' is being. To support this, Parmenides says (B8, 33): *esti gar ouk epideues; eon d'an pantos edeito* – 'for it is not lacking; yet if it were, it would lack everything'. It would lack everything; for if it lacked being in any regard, it would no longer be that which 'is'. As Parmenides states that there is no transition to being, however, it would then not be at all. Accordingly, he says (B8, 11): *houtōs ē pampan pelenai chreōn estin ē ouchi* – 'It must therefore have its essence through and through or not at all'. (Linguistically, *tou pantos edeito* can be explained by formulations such as Aeschylus, *Prometheus Bound*, 961: *pollou ge kai tou pantos elleipō* or Plato, *Phaedrus*, 235e: *tou pantos hēmartēken*). That which 'is' has now and always already reached its goal, *tetelesmenon estin*, it has now and always already been brought to completion, carried out and fulfilled. (We already encountered this word above in our discussion of Xenophanes B34.) Parmenides adds: *pantothen* – 'from all directions'. As this word is explained in l. 49, we will now already begin the explanation of the final line in this group.

Hoi gar pantothen ison homōs en peirasi kyrei: this sentence is untranslatable because it is constructed in such a way as to permit different grammatical readings, all of which make sense; bringing together the different meanings then reveals this unique line's full web of meaning, of which Parmenides says (B8, 50/1):

[13] Hermann Fränkel, *Wege und Formen des frühgriechischen Denkens* (Munich: C. H. Beck, 1955), p. 194.

en tō soi pauō piston logon ēde noēma
amphis Alētheiēs
Here shall I close for you my trustworthy speech and thought
about the truth.

The first possibility (with a comma between *ison* and *homōs*): 'For, being equal to itself from all directions, it reaches its limits uniformly' (*enkyrei peirasi*). The words 'equal to itself from all directions' are explained by B8, 29, from which we began: 'As the same within the same, it abides and rests in itself'. As it abides in itself, all directions that are possible are 'directions within being' (Fränkel); there is no outside. In itself, however, as the lines we have passed over explain, the 'existent' is equal to itself everywhere and in all directions. It is fulfilled (*tetelesmenon*) everywhere. It therefore encounters 'limits', that is, itself in its fulfilment, everywhere and in all directions. – The same meaning results if one shifts the emphasis and draws *homōs* towards *ison*: 'Uniformly equal to itself from all directions, it reaches its limits.' In this version, the first part of the sentence explains the second. – A third possibility arises if one connects *ison* to *kyrei*. In this very common construction, *kyrō* means *kyreō*: 'to be, to be situated'. Then the line means: 'For it is equal to itself from all directions, uniform in its limits.' – This leads to a fourth possibility that likewise corresponds to common usage. *pou kyreis* means 'where are you?' If we employ this meaning, the line means: 'Equal to itself from all directions, it is situated uniformly within its limits.'

If one only lists these different possibilities, there is a danger that the explanation will seem like a game. If, however, one tries to think through the different meanings individually, as if in a meditative exercise (which requires going through the whole poem in each case), and understand the agreement between them, this results in a purely noetic experience of what the One is in Parmenides; one then gains a sense of what the word *pantothen* – 'from all directions' – might mean within this One. This would be a preparation for the exercise that Plato describes Parmenides encouraging the young Socrates to undertake (*Parmenides*, 135cff.). We will be referred to the meaning of this line again later, and will have occasion to speak of a meaning that has not yet been discussed.

We now turn to the lines in which Parmenides reaches the goal of this last line, starting with l. 44/5:

to gar oute ti meizon
oute ti baioteron pelenai chreon esti tē ē tē.
For it is necessarily the case (through the binding of *ananke*) that this has its essence neither to a greater nor lesser extent here or there.

The word *pelenai*, thus *pelein*, is almost synonymous in Parmenides with *einai*, 'to be' (see B6, 8). But it has a particular weight and a particular sound, and we will therefore provisionally translate it as 'to have its essence'. In B8, 11 we read: *houtōs ē pampan pelenai chreon estin ē ouchi* – 'It must therefore have its essence [and thus be] through and through or not at all'. (In B8, 18 *pelein* does not correspond to a simple *einai*, but rather *etētymon einai*.) Given the compressed language of Parmenides, one can consider the possibility that the kinship with the word *polos* is also involved in connection with the metaphor of the sphere, 'for the ancients, unlike those who came later, did not use *polos* to mean a point and the end of the axis, but rather the all-encompassing (the firmament) as a whole' (*polon gar hoi palaioi ouch hōs hoi neōteroi sēmeion ti kai peras axonos, alla to periechon hapan*, Schol. Aristoph. Av. 179). Being cannot fulfil its essence to a greater degree here and to a lesser degree there. Thus we encounter an opposition that corresponds exactly to that between *aletheia* and *doxa*: the opposition between the eternally self-identical 'that which is' fulfilled within its 'limits', and that which has its essence to a greater or lesser degree, and comes into being or passes away. In that which 'is', which is always and everywhere maintained by the shackles of its completion, 'becoming is extinguished and destruction is lost' (*tōs genesis men apesbestai kai apystos olethros*). Lines 47/8 say the same thing:

out' eon estin hopōs eiē ken eontos
tē mallon tē d'hēsson, epei pan estin asylon.
Nor can 'that which is' be
more 'that which is' here, and less there, since it is all inviolable.

There are no differences of being in that which 'is', no rise or decline; thus it does not change over time. But where there is no change over time, there is no movement. That is why, in B1, 29, where he calls truth 'well-rounded', Parmenides refers to the 'heart' of truth as *atremes*, unshakable.

We now come to line 46/7, which we had previously passed over: *oute gar ouk eon esti, to ken pauoi min hikneisthai eis homon* – 'for there is no non-being that could prevent it from reaching the uniform.' Line B8, 22 prepares the understanding of these difficult words: *oute diaireton estin, epei pan estin homoion* – 'it is not divisible, for it is uniform as a whole' – and the related line B8, 25: *tō xyneches pan estin; eon gar eonti pelazei* – 'therefore it is cohesive as a whole; for what 'is' touches what 'is'. Behind these words is the profound thought that Zeno, the student of Parmenides, explained in his text against multiplicity. It is not possible to develop this thought here; we must restrict ourselves to the most indispensable points. A division of the One would result in several 'parts' that would have to be distinguishable from one another; for, if they were not distinguishable, the one 'part' would not be separate from the other, and the One would not have been divided at all. In the divided One, each 'part' would be 'different' from the next 'part'. Let us suppose that, in the series ABC, A were being as part of that which 'is'. Then B would have to be non-being, for only thus could it differ from A. But if it is non-being, it is not at all; thus it does not stand between A and C. Now A and C move alongside each other. If they are different, the same happens as between A and B; then C disappears. So they must be identical: *eon gar eonti pelazei*, for that which 'is' touches that which 'is'. But then we can no longer distinguish between A and C, and the whole *diairesis* is undone. The impossibility of assuming a non-existent B is noted in l. 46/7: *oute gar ouk eon esti, to ken pauoi min hikneisthai eis homon.* Now we can translate this more accurately: 'for nothing non-being is possible that could interrupt it in reaching its sameness'. The words *hikneisthai eis homon*, 'reaching its sameness' mean the same as the words *eon gar eonti pelazei* – 'for the 'that which is' touches 'that which is'': in that which 'is', the different does not reach the different and the different does not lie alongside the different; rather, it is indivisible sameness. This is indicated by the words *epei pan estin homoion*, which we can now translate: 'as all is the same'.[14] These verses too are an interpretation of B8, 29: *tauton d'en tautō mimnei kath' heauto te keitai* – 'As the same, it abides in the same and rests with itself.' This insight explains why *to eon*, 'that which is', cannot refer in Parmenides to one 'that which is' alongside another 'that which is' – not even in the sense of Platonic 'ideas' – but rather to the whole, whole in its sameness, of the 'well-rounded' truth of being. (Concerning *pan*, see in B8 5, 22, 24, 25, 28; based on these, the interpretation of *pantothen* in B8, 43 and 49 would have to be undertaken again. *pantē* in B8, 44 also belongs in this context.)

Thus understood, the abiding of the same in the same is articulated by Parmenides in B8, 25 with the words: *tō xyneches pan estin* – 'therefore it is cohesive as a whole.' Through the mediation of Zeno, this concept became the concept of the continuum through which, in sharp contrast to Parmenides, we ontologically constitute 'space' and 'time'. We must therefore go a step further to see how this ontology shows itself in Parmenides's treatment.

We have hypothetically supposed the divisibility of the One, and have seen that the 'parts' of the divided One must then be distinguished. Parmenides shows the various ways in which such an investigation could be attempted.

One could distinguish – in 'space', we would say – according to places (*topon allassein* in B8, 41). If we also keep in mind the insight we have gained that being is cohesive in itself, one must assume that the places form a continuum. This continuum is verified in the movement that passes through the 'places' in the continuum in succession. The result of this is described by Zeno (B4) in the words: *to kinoumenon out' en hō esti topō kineitai out' en hō mē estin* – 'the moved moves neither in the place where it is nor in the place where it is not.' Let us suppose that the series ABC represents the 'places'. If we say that the moved is now in place B, we have brought it into a position and thus undone the movement. If we say that it is in transition from A to B, there are two ways of interpreting this statement: either we view AB as a distance – then this distance can be divided into an infinite number of points, i.e. places, and the aforementioned conditions apply to each of these positions. Or we reject this divisibility and say that indivisibility, according to Parmenides, constitutes the nature of the continuum. Then there are again two possibilities: either we maintain that the moved *is* if it is in a place. This would mean that in a placeless transition from it, one can no longer say that it is; but if it is not, it cannot move. Or we abandon the being of the places and say that the being of the moved is a 'being-in-transition'; thus we have once again transferred being into the unity of the continuum. But if the difference between places is not, then movement is no longer movement, for movement was *topon allassein*, change from place to place. This

[14] Concerning *homoion* as 'completely the same', see Wolfgang Kullmann, 'Zenon und die Lehre des Parmenides', in *Hermes* 86, p. 168.

means: continuum and difference of place are irreconcilable, for difference as such, as we saw, is discontinuity. Either it is the continuum or it is difference. But difference undoes itself, for being different would only be possible as an alternation between being and non-being. Thus only the continuum *is*. But a continuum in which there is no difference between places is not 'space'; thus being is not in possible places, not in space. If we nonetheless maintain, based on our experience, that something like movement, the moved and difference of places *is*, then we are like the indistinct masses *hois to pelein te kai ouk einai tauton nenomistai / kou tauton* – 'for whom being and non-being are the same and not the same' (B6, 8/9). Then we have gone astray in the confusion of *doxa*.

What is true of difference of place also applies to other aspects that could be used to make a distinction in being: difference of time (B8, 5), of intensity (B8, 24/5 etc.) or of qualities (B8, 41). Being abides in itself as the same. According to the untranslatable line (B8, 49) discussed above: *hoi gar pantothen ison homōs en peirasi kyrei*. Because of the insight we have now gained, however, we must revise the translation of *homōs* given above. We used the word 'uniform'; but the Greek word for 'uniform' is *homalos*, whereas *homos* means 'one and the same'. Parmenides therefore uses the word *homōs* like an adverb for *hen*, 'the One'. *homōs* means 'in the manner of the One', 'in the One'; it is explained by the words in B8, 5: *epei nyn estin homou pan* – 'because it [the 'existent'] now *is* at once as a whole.' Therefore, we will henceforth translate *homōs* as 'at once and in the One'.

Change of place, change of intensity and change of qualities: these are three different forms of movement. All movement is change over time. All other definitions are therefore a consequence of the fundamental definition of that which 'is' by which it is removed from change over time: *to ge Moir' epedēsen oulon akinēton t'emenai* – 'fate has bound it so as to be whole and immovable' (B8, 37/8). We already encountered the conveyance of that which 'is' away from change over time on the way from Homer to Xenophanes, in the being of the 'ever-being' gods. Through it, we are referred once again to Hesiod. We already discussed the lines from the prooimon of the *Theogony* (32f.):

> but they breathed song into me,
> divinely heralding song, so that I would praise what will be and what has been,
> and told me to sing of the race of the blessed, of the ever-being.

We saw how the Homeric 'eternity formula' adopted again in l. 38 is altered: what will be in the future and what formerly was, as the transient that changes over time, are here set against the ever-present being of the gods and both of these, the eternal and the transient, are to be sung of by the poet. This is the task set for the poet Hesiod by the Muses. This corresponds exactly, in an altered sequence, to what the goddess proclaims to Parmenides (B1, 28ff.):

> It is necessary for you to learn everything,
> both the unshakable heart of well-rounded truth
> and the opinions of mortals, in which there is no true reliability.

Truth is in the place of the ever-being race of the blessed, while the views of mortals are in the place of what is perceived in them: change over time. So the distinction between *alētheia* and *doxa* emerged from the basic experience among the Greeks of the opposition between gods and humans, between immortality and mortality and between the eternal presence of being and the alternation of becoming and passing away in time. As we are seeking the foundation of the distinction between *nous* and *doxa*, we will have to investigate further the relationship between being and time in Parmenides.

Parmenides defines the relationship between being and time in B8, 5 with the words *oude pot' ēn oud' estai, epei nyn estin homou pan* – 'And it never was, it never will be, for it Is now as a whole.' To understand these words, one must set them against the being of the 'ever-existent' gods. The gods are 'always': that means they are now, as they already were in the past, and as they will be in the future. Their ever-being is a constancy that continues throughout change over time without being touched by it. The 'that which is' of Parmenides, by contrast, is not 'always' but 'now'; it has no extension in time whatsoever, and he therefore notes: 'it never was, it never will be'. Parmenides was careful to avoid the word *aei* – 'always' – in the part dealing with *alētheia*, even though he speaks throughout of the motionlessness of that which 'is'; it only appears in the part on *doxa*, where the moon is described (B15) thus: *aiei paptainousa pros augas ēelioio* – 'always watching out for the rays of the sun'. But here the 'always' has the same necessity as the 'now' from the line quoted above. Because the 'now' only becomes comprehensible through its contrast to the 'always', we will first of all examine the 'always'.

Line B15 has its model in Book V of the *Odyssey* (l. 274). On his journey on the raft, Odysseus looks to the Pleiades and the late-setting Bootes and the Bear, also known as the Wain, *hē d'autou strephetai kai t'Ōriōna dokeuei, oiē d'ammoros esti loetrōn Ōkeanoio* – 'which ever circles in its position and gazes watchfully at Orion, and alone has no part in the baths in the Ocean.' In both passages, the 'watching' concerns the relationship between two stars that stand in an 'aspect' to each other and thus form a constellation.[15] In Parmenides, however, the moon watching out for the rays of the sun has the special meaning that the moon, as the Greeks knew since Thales, receives its light from the sun's rays. It is *nyktiphaes peri gaian alōmenon allotrion phōs*, 'shining by night with borrowed light, wandering around the earth' (B14), and thus represents the mixture of light and night that, according to Parmenides, defines the nature of the sublunar sphere of *doxa*.

The connection between the constellations of stars watching out for one another and the eternal course of time is described in a fragment by Plato's uncle Critias (B18):

akamas te chronos peri t'aenaō
rheumati plērēs phoita tiktōn
autos heauton, didymoi d'arktoi

---

[15] Since *horan* or *blepein* is a technical term from astrology, as shown by Franz Boll in *Kleine Schriften zur Sternkunde des Altertums* (Leipzig: Koehler & Amelang, 1950), p. 121, the word may refer to Babylonian astronomy.

tais ōkyplanois pterygōn rhipais
ton Atlanteion tērousi polon.
Time, tireless and filled with
ever-flowing streams, goes on its orbit,
birthing itself, and the twin bears
with the swift movement of their wings

Here, time is the eternal rotation of the firmament. But Atlas, who carries the celestial pole and thus the entire firmament, and around whom day and night and the procession of countless stars revolve, is for Critias, as we learn from Fragment B19, the divine *nous* on the model of Anaxagoras. Parmenides too understood the divine 'always' as the eternal orbit of the stars. The goddess proclaims to him (B10, 4ff.):

erga te kyklōpos peusē periphoita selēnēs
kai physin, eidēseis de kai ouranon amphis echonta
enthen [men gar] ephy te kai hōs min agous' epedēsen Anankē
peirat' echein astrōn.
And you will learn of the wandering deeds of the round-eyed moon
and of her nature, and you will know the heavens that surround us,
whence they arose, and how necessity took them and bound them
to keep the limits of the stars.

The rotation of the sky, driven by necessity, takes the stars in their spheres along with it; that is why the moon 'always' watches out for the rays of the sun. But this orbit, as a movement, is a connection of rising and setting, of day and night, of becoming and passing away; it is being and non-being in one, and therefore *doxa* – semblance. At the same time, however, we learn from this passage how this *doxa* relates to being. For the description of the firmament refers verbatim to lines 30/1 of B8, which explain the self-abiding of the 'existent':

choutōs empedon authi menei; kraterē gar Anankē
peiratos en desmoisin echei, to min amphis eergei
And thus it will abide constantly there [and not move like the firmament]; for mighty necessity keeps it in the shackles of the boundary that encompasses it.

In B8, 37 and B10, 6, Parmenides uses *epedēsen* rather than *en desmoisin echei.* Just as the shackles of Ananke enclose that which 'is' in the boundary of its fulfilment, so too the sky encloses the stars in their boundaries with the shackles of Ananke. The nature of these boundaries emerges in the claims that the sky is at once driven (in motion) and bound (at rest). For this combination of movement and being bound in the One is only possible in the regular rotation of a sphere. If movement is to occur at all, the rotation of the sphere is the highest approximation to being's abidance in the One.

The equivalent wording of the two passages thus points to an inner connection. According to B8, 60, the sphere of *doxa* is an *eoikōs diakosmos* – 'an orderly arrangement [of light and night] that is similar.' We already encountered the concept of the ‘similar’ (*eoikos*) in Xenophanes B35: *tauta dedoxasthō men eoikota tois etymoisin* – ‘this shall be presented as a *doxa*; it resembles the genuine.’ We already saw in the shift away from Hesiod in these lines that, in relation to being, a semblance that looks genuine is not like a deception, but rather appearance – an appearance that soon becomes deception, admittedly, if one views it not as mere appearance but as being. It is therefore vital to distinguish relentlessly between being and non-being, that is, between being and appearance – not only for a recognition of the truth, but also for the recognition of appearance as such, and thus for an appropriate comprehension of this world of ours as it is.

But the equivalence, based on similarity, between the firmament and that which ‘is’ which rests as itself within itself, also comes to light in a more important passage in the middle of the whole poem. In the passage we have yet to explain, B8, 43, Parmenides says that the being which is brought to fulfilment from all directions is *eukyklou sphairēs enalinkion onkō* – ‘comparable to the mass of a well-rounded sphere’. Only now does the deeper meaning of this comparison become visible: the sphere is not an arbitrarily chosen image, but the sphere of the firmament. And that which ‘is’ is ‘comparable’ to this sphere because it is brought to appearance in this sphere. For only the true appearance of ‘that which is’ can be a ‘comparison’ that is adequate to that which ‘is’. The correspondence concerns not only the boundaries, but also the mass of the sphere. B9, 3 states of the celestial sphere’s content that *pan pleon estin homou phaeos kai nyktos aphantou* – ‘everything is filled at once with light and invisible night’. In B8, 24 there is an equivalent passage: *pan empleon estin eontos* – ‘it is entirely fulfilled by that which ‘is’’. And the *homou* – ‘at once’ in the eternal rotation of the firmament (B9, 3), corresponds to the words in B8, 5: *epei nyn estin homou pan* – ‘for now it Is at once as a whole.’

But this word leads us back to that which ‘is’ itself, for which the sphere of the firmament and the constancy of the Always that appears in its rotation, which in its essence appears in the passing of time, was a metaphor. We now see that the word *homou* – ‘at once’ – in its combination with the Now indicates a contrast to the infinite extension of the ‘Always’ through the past and the future. But how should this contrast be understood? How can being be as a whole and now ‘at once’? Earlier on, proceeding from the invocation of the Muses at the start of the Catalogue of Ships, we followed the development in Greek thought of the perception of the divine *nous*, for which everything is present at once. *Panta idōn Dios ophthalmos kai panta noēsas* – ‘the all-seeing and all-knowing eye of Zeus’ (Hesiod, *Works and Days*, 267) sees everything, like the Homeric Muses, in the immediate present as an eyewitness. Its knowing is by its nature parousia, omnipresence (see *Iliad* II, 485: *pareste te iste te panta* – ‘you are present and know everything as eyewitnesses’). But the knowing of the *nous* not only has the spatially separate in its view at once; like the knowledge of Homer’s seer (*Iliad* I, 70), its immediate seeing encompasses *ta t’eonta, ta t’essomena pro t’eonta* (‘what is, what will be and what has been’). For the future, it is immediately clear that a knowledge which is understood as direct perception, as being an eyewitness, is only possible for future events if these are already present before the *nous*. The diverging extension in time is therefore, if a having-in-view of future events is to be possible, that semblance which is uncovered in the ‘Zenonian’ paradoxes originating from Parmenides. But whoever undoes the future also undoes the past. Nor is the divine having-in-view of the past the remembering self-movement here and there that, according to *Iliad* XV, 80ff., is a property of the human *nous* and serves as a comparison with the flight of the gods. For that remembering self-movement is only possible because the area in which the memory moves here and there has already been disclosed in its unity and openness; it is conditioned by the disclosure of this area. But here the concern is being and truth itself: not memory but the condition of possibility of memory. Not the knowledge of given facts, but rather the original disclosure of the area in which human knowledge and the assertion of knowable facts in their manifold appearances become possible in the first place. The eye of the *nous* for which this area is originally disclosed is not forced to pursue the appearances scattered among the variety of times and places in this area once they have been given. Because this area is open to it in its unity and as a whole at once, it is an eyewitness to everything in the One.

These remarks may be useful as a preparation for the words of Parmenides to which we will now turn. We begin with the two first lines of Fragment B4:

leusse d'homōs apeonta noō pareonta bebaiōs;
ou gar apotmēxei to eon tou eontos echesthai.
Behold how the absent is securely present for the *nous* at once and in one;
for it will not cut off that which 'is' from holding on to that which 'is'.

In order to understand this fragment, we must first do away with a linguistic misunderstanding. Instead of *homōs* – 'at once' – the fifth edition of the pre-Socratic fragments contains *hómōs* – 'nonetheless'. The translation reads 'But behold...' This presupposes a preceding sentence to which the one beginning with 'behold' stands in contrast. The accentuation is not a question of textual authority but of interpretation. It is only justified to decide on an interpretation with an additional hypothesis about the preceding sentence if the words are not comprehensible in themselves. If one wishes to explain the words on their own terms, however, one must attempt this with *homōs*.

To explain this line, we must assume that the following line with *gar* ('for') gives the reason. The words *ou gar apotmēxei to eon tou eontos echesthai* – 'for it [the *nous*] will not cut off that which 'is' from holding on to that which 'is'' – correspond to the now-familiar words from B8, 22-25: *oude diaireton estin [...] tō xyneches pan estin* – 'And it is not divisible [...] therefore it is cohesive as a whole'. The explanation for the indivisibility and continuity in the sameness of being is given in B8, 46/7: *oute gar ouk eon esti, to ken pauoi min hikneisthai eis homon* – 'for nothing non-being is possible that could interrupt it in reaching its sameness [*homon*]'. Likewise, the statement in B8, 22 that *oute diaireton estin* – 'and it is not divisible' – is explained by the words *epei pan estin homoion* – 'for everything is the same [like the *homon*]'. Regarding the 'cohesion' of the 'existent', the word also appears in B8, 5/6: *epei nyn estin homou pan / hen, syneches* – 'as it is now 'at once' as a whole, One, cohesive.' And all these aspects are summarized in the 'untranslatable' line B8, 49: *hoi gar pantothen ison homōs en peirasi kyrei.* In the light of all this, there can be no doubt that we should read *homōs* also in B4.

Leusse – 'behold': to understand this word, we must consider that the prooimion, which describes the poet's removal from the 'house of night' and into the light, follows on in its language and images from the initiation rites of the mysteries that flourished in Lower Italy in connection with the belief in the hereafter.[16] The highest rite in the mysteries is that of *epopteia*, 'beholding'. In the course of initiation into the 'mystery of being' (Jaeger), the line introduced by *leusse* – 'behold' – can only have been placed where the epopteia first anticipated in the prooimion actually takes place and the epiphany of being reveals itself to the poet, namely at the start of the part of the poem referred to in B8, 50/1 as *pistos logos ēde noēma amphis Alētheiēs* – 'trustworthy speech and thought about the truth'. Correspondingly, the section on *doxa* begins in B8, 52 with the appropriate command for the new area: *manthane* – 'learn about'.

In the Greek, the verbs of seeing are followed by the participle; hence the correct translation of the lines would be: 'Behold how, at once and in one [*homōs*], the absent is securely present for the *nous*'. If everything is present at once and in one, the absence of the absent is undone: *oude pot' ēn oud' estai, epei nyn estin homou pan* – 'and it never was, it never will be, for it Is now as a whole' (B8, 5). Lines B4, 1 and B8, 5 are congruent. Hence the *apeonta* – 'the absent' – is to be understood as the temporal in *physis*; it is *ta t' essomena pro t' eonta* – 'what will be and what has been'. Before being reveals itself in epopteia as the One present of the One, its human meaning is still tied to the course of time; the goddess's proclamation refers to the future (B1, 31 *mathēseai* – 'you will learn about'; with the accumulation of future forms, it is worth considering whether B10 belongs in the vicinity of B1, 31/2). This temporality is now undone; the absent is present, and absence is subsumed by the presentness of the 'now'. But this also undoes the distinction between human and divine *nous*; as soon as the word *leusse* – 'behold' – is uttered, the poet's *nous* is subsumed by the One *nous*. Hence there are two equally valid interpretations of this line. One can also form the connection *leusse noō*: 'behold with the *nous*'. Just as the diversity of appearances is subsumed by the One truth of being, the diversities of linguistic form are subsumed by the unity of an unchanging 'meaning'; thus the same sentence can permit several 'constructions' that are all both different and correct and mean the same even in their difference, and the interpretation of the different lines therefore shows time and again that, strictly speaking, they are saying the same thing. But this linguistic phenomenon has its origin in the event that takes place with the word *leusse* – 'behold': in the union of the human and the divine *nous*. The beholding that the poet is authorized to undertake is therefore the beholding by the eye which Hesiod (*Works and Days*, 267) called *panta idōn Dios ophthalmos kai panta noēsas* – 'the all-seeing and all-knowing eye of Zeus'. And so what Homer says of the Muses in his invocation of them in Book II of the *Iliad* directly fulfils itself in the poet himself: *pareste te iste te panta* – 'you are present and know everything as eyewitnesses'. Now it is the poet himself for whom everything is present at once and in the One; he has, as Parmenides says in allusion to the language of the mysteries,

---

[16] See Werner Jaeger, *The Theology of the Early Greek Philosophers* (Oxford and New York: Oxford University Press, 1948).

become one who 'has seen'; unlike the mortals who have seen nothing and know nothing (B6, 4ff.), he is a 'knowing man' (B1, 3).

The invocation of the Muses also explains the word *bebaiōs* – 'securely'. Speaking of human knowledge that is not based on the presence of the eyewitness, Homer states: *hēmeis de kleos oion akoumen oude ti idmen* – 'but we only hear the news and know nothing'. This recognition of the deceptive indirectness of all human knowledge is, as we have seen, also the basis of Fragment B34 by Xenophanes:

> No man has ever seen what is immediately clear, nor will there ever be one who [as an eyewitness]
> would have knowledge about the gods and of what I say about all things;
> for even if it occurred to the fullest extent that someone spoke in a concluding fashion,
> he would not know it from his own observation, for a semblance is created over everything.

Only against the background of this fragment by Xenophanes does one grasp what happens in the beholding of Parmenides B4, 1: the eye beholds what is clear, of which Xenophanes says that no one has ever seen it nor ever will; its path leaves the realm entered by humans (B1, 27)[17] and therefore that which 'is' is present for it, without the mediation of the Muses, in its unity with the divine *nous* – securely, not as appearance, but in its immediate being.

In Parmenides, the certainty to which the word *bebaiōs* – 'securely' – refers is called *pistis alēthēs* – 'true reliability'. It is not a certainty *about* the truth, but rather immediate truth itself, that is, the reliable presence of being. That is why Parmenides can say (B8, 12/3):

> *oude pot' ek mē eontos ephēsei pistios ischys*
> *gignesthai ti par' auto.*
> and never will the power of reliability allow that anything beside itself
> should come about from non-being.

This corresponds exactly to B8, 27/8:

> *epei genesis kai olethros*
> *tēle mal' eplanchthēsan , apōse de pistis alēthēs.*
> [...] since becoming and passing away
> were driven far astray [i.e. into the realm of *doxa*]; but true reliability cast them away.

In both passages, we too are 'driven far astray' if we follow modern ways of thinking and state: 'The certainty of the correct doctrine does not permit the claim that anything could come about from something that is not.' For with such certainty about that which 'is', thinking would be something else in relation to that which 'is'; as something else, it would itself be non-being, and as non-being it would be untrue. So its certainty would not be true, it would not be a *pistis alēthēs*, but would fall under the *brotōn doxai, tais ouk eni pistis alēthēs* – 'opinions of mortals, in which there is no true reliability' (B1, 30). Persuasion (*peithō*) directly follows *Alētheia* (B2, 4),

[17] See Kullmann, 'Zenon und die Lehre des Parmenides', pp. 159f.

for it is that which 'is' itself in which it is conveyed (B8, 35). Here too, non-being gains no admittance; the power (*ischys*, B8, 11) of *pistis alēthēs* lies in the fact that it is identical to truth itself, that is, to the presence of being.

Once the *nous* of the poet has become one with the divine *nous* through epopteia, the presence of being is revealed to him at once and in One. One can therefore take *homōs* in B4, 1 together with *leusse* – 'behold': 'behold at once and in One.' Fragment B5 leads us in this 'direction within being':

> *xynon de moi estin.*
>
> *hoppothen arxōmai, tothi g'ar palin hixomai authis.*
>
> 'It is common with me; wherever I begin, I shall surely arrive there again.'

This fragment is not from the prooimion but from the part in which the goddess speaks; she (and not the 'proud self-awareness' of Parmenides) is the I (*egōn*, B2, 1), which can say: *xynon moi estin* – 'it' (the 'existent') 'Is with me' (Aletheia) 'common'. *Xynos* is the epic and Ionic form of *koinos* – 'common, shared'; as Heraclitus shows, it was one of the highest words in the thought of the time. In Heraclitus B114 we find a profound play on words that connects the words *xyn nō* ('together with the *nous*') to *xynō*. Perhaps there is a similar play on words here: for Parmenides, the common, the *xynon* (first syllable long), is that which is together with truth in the One, namely the *xynon* (first syllable short).[18] Either way, the words mean that for *Alētheia*, truth, the beginning and the end of which the next line speaks lie together in the shared One. So they are not separate, as in the temporal divergence of the *doxa*, they are not *apeonta* – 'absent in relation to each other' – but *homōs pareonta* – 'present at once and in One'. Understood in this way, the fragment says the same as the 'untranslatable' line B8, 49 if one also connects *hoi* to *ison* and *kyrei*: *hoi gar pantothen ison homōs en peirasi kyrei.* We can now paraphrase and translate it thus: 'for, equal to itself from all directions, it encounters itself from all directions at once and in One within the limits of its fulfilled present.'

We previously related the word *leusse* – 'behold' – from Fragment B4, 1 to epopteia as the highest rite in the mystery cults. For this fragment too, tradition allows us to draw a connection to epopteia. In a fragment that is connected to the Eleusinian mysteries by our source, Pindar says (fr. 137 S.):

> *olbios hostis idōn kein' eis' hypo chthon';*
>
> *oide men biou teleutan,*
>
> *oiden de diosdoton archan.*
>
> Blessed is he who has seen this and then goes beneath the earth;
>
> he knows the end of life,
>
> he knows the beginning given by Zeus.

The initiate of whom these lines speak was an *idōn*, one who had seen. He therefore knows, for he knows as an eyewitness. What he has seen and knows are both in one:

[18] It is unclear whether the syllable *xyn-* was the end of a spondee in the incomplete line. If so, as I would assume, then the dual interpretation is possible, for the lengthening of the *n* is acceptable.

the end and the beginning of life. This One that unites the two is given by Zeus – Parmenides would say, by the *nous*. But he who has seen it is blessed if he goes under the earth: for then he is in true immortality. – We do not know what was beheld in the epopteia of the Eleusinian mysteries,[19] but Pindar's lines tell us enough to draw a connection to the epiphany that takes place before our eyes in the speech of the goddess. Through its union with the divine *nous*, the poet's *nous* likewise stands in the eternal presence of being. 'Becoming is extinguished and passing away is lost' (B8, 21). Now that everything is at once in the present, he is immortal. As a 'knowing man', that is, as an initiate, he is 'invulnerable' (*asinēs*, B1, 3),[20] for the presence of the being that he has entered is 'inviolable' (*pan estin asylon*, B8, 48). The time in which the thing we call life becomes and passes away has revealed itself as *doxa*, as mere appearance. It is the combination of being and non-being, of light and night, in which the opinions of mortals are active. *Doxa* and mortality are firmly connected in Parmenides. This is against that which 'is', which is the same as *noein* – unbecome, everlasting, so far removed from the emergence and passing away of mortals that even the ever-being of the stars is still a semblance next to its presence in the 'now'. It is in this 'now' that *Alētheia* stands when she says:

> It is common with me; wherever I begin, I shall surely arrive there again.

But do the references to 'beginning' and 'returning' not suggest a path that leads from a point of departure to a goal? Parmenides indeed speaks of a path that follows *Alētheia*: it is the path of persuasion (*peithous keleuthos*, B2, 4) on which the words of the goddess lead the poet. So this path is identical to that speech: it *is* the poem whose fragments we are permitted to read. If this path is meant to be a path of truth, it must abide in the presence of being. The terrain it traverses is described exhaustively by Parmenides in B2, 3 with the words: *hōs estin te kai hōs ouk estin mē einai* – 'it is called: 'Is', and there is no 'Is-not'.' We thus remain constrained by *doxa* if we read the poem as a series of lines with a 'train of thought' that leads to a 'result'. Rather, the poem must be read in such a way that every part says the same thing. Only thus does it abide in the sameness of being, and only thus is being conveyed in that which 'is' itself (B8, 35f.):

> For you will not find thinking apart from that which 'is'
> in which it is uttered.

Therefore in each individual piece there is a departure and a return to this point of departure of which Fragment B5 speaks. We have tried to retrace this circling movement in the One in our explanation.

Retracing this movement in Fragment B5 has not yet brought us back to the starting point, however, namely the word *xynon* – 'common'. Aletheia says: *xynon de moi estin* – 'it is common *with* me'.

---

[19] For conjectures based on finds from Lower Italy, see Nilsson, *Geschichte der griechischen Religion*, vol. 1, pp. 639f.

[20] I consider Meineke's reconstruction of the text reliable, also with regard to the discussion of this question in the commentary by Leonardo Tarán in *Parmenides* (Princeton University Press, 1965). See Jaeger, *The Theology of the Early Greek Philosophers*, p. 98n23.

We have so far, and rightly, taken this dative to mean that that which 'is' is '*for* me', 'in relation to me', 'in my eyes', something shared in the One. This interpretation is correct. But it is equally correct and necessary to consider the other interpretation, which sees that which 'is' as a common One *with* the truth. And this means that the *logos* of the goddess must also be One *with* 'that which is' and the truth, for the word *xynon*, as we saw, refers to the beginning and return of this *logos*. The *xynon* is what Parmenides calls the One. Now this One must be understood as the unity of that which 'is', of *Alētheia* and of the epiphany of *Alētheia* in the *logos* of the goddess.

The question of the unity of that which 'is' is the question of its 'cohesion'; for in Parmenides, the *hen*, the One, is the *syneches*, the cohesive (B8, 6). We have already developed the concept of *syneches*, based on Zeno's paradoxes of movement, from the perspective of formal unity, without considering that the *syn*, the together, seems to point to something like a plurality within this unity. And yet we cannot pass over this question; for, coming from Xenophanes, we entered the poem of Parmenides with the question of what the 'same' is in which the divine *nous* abides, and we saw that Parmenides views this same as sameness within a difference, as the same in *noein* and *einai*. Is the 'cohesive' element that which holds *noein* and *einai* together? Is the *xynon* the bracket of sameness in which the dual abides as the One?

This question forces us to return to line B8, 34: *tauton esti noein te kai houneken esti noēma* – 'Thinking and of what is the thought are the same.' In these lines, *noein*, as we have followed from Homer onwards, means the seeing of the 'divine' *nous* as the direct having-in-view of the omnipresent 'whole' of that which 'is'. As 'thinking' can involuntarily suggest having in mind something previously unthought, it would be better to replace it with 'beholding', in the sense of the direct omnipresence of the 'divine' *nous*. Then the sentence would read: 'Beholding and that for which beholding Is are the same.' If we are to penetrate the mystery of this sentence, we must start from the linguistic form. The words *te kai* – 'both and' – are used to join two components into a whole; these are evidently the components that are the same thing. The first component appears to be simple: *noein* – 'to behold'. But the second component creates difficulties: *houneken esti noēma* – 'that for which beholding Is'. The next sentence explains: *ou gar aneu tou eontos [...] heurēseis to noein* – 'For you will not find beholding apart from that which 'is''. We therefore hastily concluded that *houneken esti noēma* – 'that for which beholding Is' – is the *eon*, that which 'is'. But does it seem plausible that Parmenides would, in such a strictly formed text, have been so indifferent in placing a participle – *to eon*, that which 'is' – alongside an infinitive – *noein*, 'to behold' – as its equivalent? In Fragment B3, which we have yet to discuss, he says: *to gar auto noein estin te kai einai* – 'for it Is the same to behold and to be'. We will have to assume that the two lines are congruent; then the word *houneken* – 'of what' – in B8, 34 would stand for the word *einai* in B3. The words *houneken esti noēma* would then mean that that which is beheld Is because of the infinitive form of being, *einai*. So why does Parmenides not simply say *einai*, 'to be', here too, as in B3? What is the purpose of *noēma*, which joins the two infinitives we have now found as a third element, a neuter noun?

With the word *noēma*, sameness returns as if in a circle to the *noein* from which it had begun. But it has passed through *einai* and absorbed it. Just as *thauma* refers to the wonder at which one wonders and *mathēma* what is learnt through learning, so too *noēma* is that which is beheld in beholding. The grammarians call such a construction a 'nomen rei actae'; the act indicated by the verb appears completed in the noun that is formed from it. Parmenides would say: *tetelesmenon estin – noein* has reached its goal in the *noēma*, has been completed and fulfilled; the *noēma* is the *peiras*, the 'limit' of *noein*. A *noein* that does not reach the *noēma* is beholding that does not behold, a blind gaze. But a beholding that does not behold is not a beholding, and because it is not a beholding, it is not at all.

But the line does not say that the *noēma* is because of *noein*; rather, it contrasts *noein* with that because of which the *noēma* Is. The *noēma* Is because of *einai*, the infinitive form of being. In Greek thinking, the *noēma* comes from that which appears in what is beheld. That which appears to the beholding eye in that which is beheld upon is being. It is, as the comparison with B3 showed, not the *eon*, 'that which is', but infinitive being. Yet what does this mean? If we are to understand sameness as the sameness of two coordinated, equally significant and corresponding components, as is demanded linguistically by *te kai* – 'both and' – then the two infinitives must stand in the same relation to the neuter noun *noēma*, in which they meet. So *einai* must stand in the same relation to the *noēma* as does *noein*. For *noein*, the *noēma* was a *nomen rei actae*. Can there also be a *nomen rei actae* for einai? The *nomen rei actae* for *einai* would have to refer to that in which being 'reaches its goal, is completed and fulfilled'; it would have to be that in which being reaches its *peiras*, its 'limit'. Being reaches its *peiras* in the *eon*, that which 'is'; thus the *noēma* Is the *eon*. That is why Parmenides can continue: *ou gar aneu tou eontos [...] heurēseis to noein* – 'For you will not find thinking without that which 'is''. This is the same as saying: *ou gar aneu tou noēmatos heurēseis to noein* – 'for you will not find beholding without that which is beheld'. But how can being find its completion and fulfilment in the *noēma*, that which is beheld? Should one not think that being per se *is*, whether it is beheld or not?

If we ask these questions, our mortal thinking has already separated the two components, which are one. Then they stand beside each other as different, and each of them, because it is different, is a non-existent in relation to the other. Hence separation undoes being as such. So *einai* depends as much on *noein* as *noein* on *einai*. Completed being, the *eon*, *must* be the *noēma*, just as completed *noein*, the *noēma*, must be the *eon*. Viewed in this way, however, the bracketing remains formal. Thought from the perspective of being, its dependence on *noein* comes from the fact that being, as we have followed from Homer onwards, is understood as *par-einai*, as being-present before the eye of the *nous*, just as the *nous* is conversely understood as the present that sees everything. An absent being would either not be at all – for then it is not being – or it would not yet or no longer be. Yet whatever is not yet or no longer, Is not; being is therefore kept within the 'limits' of being-present. But the present in which being Is, is its being-revealed, and it can only be revealed for *noein*. This means that completed being, the *eon*, that which 'is', must by its nature be *noēma*. The *noēma* Is completed beholding and completed being. It Is the same thing that *noein te kai einai* Is.

It may seem as if we had now reached our goal, namely the 'limits' in which beholding and being fulfil themselves at once and in the One. But this is not yet the end, for the clarity we have reached is blurred in the next line if we consider the words we have previously omitted:

For you will not find beholding apart from that which 'is'
in which it is uttered.

What does uttering mean here? Who is it that speaks? And how does speaking integrate itself into the sameness of that which 'is', which Is at once and in the One the completed cohesion of beholding and being?

There is another passage in which the *noēma* is connected to speech. At the end of the first part of the poem (B8, 50/1), we read:

en tō soi pauō piston logon ēde noēma
amphis Alētheiēs
With this shall I close for you my trustworthy speech and the *noēma*,
which is about the Truth.

Walther Kranz translates: 'my trustworthy speech and thought about the truth' [*mein verlässliches Reden und Denken über die Wahrheit*]. This would be correct if the speaker were the mortal human Parmenides, concluding a discourse on the specified subject of truth. But the speaker is the goddess *Alētheia* herself, and *Alētheia* is the truth of being. Her *logos*, as we have seen, is her epiphany, the epiphany of the truth of being. If that which emerges in this logos as a 'teaching' is not to be spoken into the wind as a meaningless *doxa*, this teaching must tell the truth about *Alētheia* herself and her epiphany. The lines refer to a *logos amphis Alētheiēs* – a *logos* that is about *Alētheia*. If one separates the *logos* from *Alētheia*, then both – *Alētheia* and the *logos*, fall into non-being. But how can both of them *be* in keeping with this *logos*? We are told this by the words *ēde noēma* – 'and that which is beheld'. It is inconceivable that the word *noēma* should mean something different here from what it meant 15 lines earlier. But if it means the same thing, this brings to light a thought that is outrageous and almost intolerable for our mortal thinking. We saw that the *noēma* Is the *eon*, that which 'is', as the 'whole' of the 'well-rounded' truth of being. Now we hear that the *noēma* is the 'trustworthy speech' of the goddess. So we must conclude that this speech, namely the poem whose fragments we are reading – this trustworthy speech Is the *eon*, the 'whole of the well-rounded truth of being'. We want to resist, but this exact thought is necessary. For the *logos*, as we saw, is the epiphany of *Alētheia* herself. But what is the epiphany of truth meant to be, if not that which 'is' whose truth it Is?

To reassure ourselves of the ground under our feet at this disturbing point on our path, however, we consider another line in which *legein* is connected to *noein*. It is the first line of Fragment B6: *Chrē to legein te noein t'eon emmenai; esti gar einai*. Once again, this line is untranslatable because it permits different equally valid constructions:

1. 'It is necessary to say and think that (only) that which is, is; for being is' [*Nötig ist zu sagen und zu denken, dass nur das Seiende ist*; *denn Sein ist*] (Kranz). In this reading, *eon emmenai* ('that that which is, is') is an infinitive construction that depends on both *legein* and *noein* and indicates what it necessary 'to say and to think'. Based on our findings, we must view the *eon* as that which 'is', that is, as the whole of the well-rounded truth of being. In the speech of *Alētheia*, *legein* must not be understood as 'mortal' speech and *noein* must not be understood as 'mortal' thinking; this *legein* is the revelation of the truth of being in epiphany, and this *noein* is a beholding of revealed being as *eon*, as 'that which is', in its present. If one reads the words in this way, however, it is 'necessary' – as a necessity of Ananke, Dike and Moira – that this *legein*, bound by the shackles of Ananke, *cannot* and must not reveal anything other than the fact that that which 'is' Is. And it is equally necessary that the beholding for which 'that which is' is present should also behold nothing other than this: that 'that which is' Is.
2. 'It is required that an utterance and thinking of what Is, Is; for being exists' [*Es ist erforderlich, dass ein Aussagen und Denken dessen, was Ist, Ist*; *denn ein Sein gibt es*] (Fränkel). Here *to legein te noein te* is the subject of the infinitive *emmenai*, and *eon*, as the object of *legein to noein te*, refers to what is uttered and recognized. The translation of the second clause should be: 'for being Is'; that which 'exists' stands alongside something else that also 'exists'; together with its being-respectively-different, it also carries non-being as such and therefore belongs in the realm of appearance, of *doxa*. 'Required' should be replaced with 'necessary'; for it is the power of Ananke, Dike and Moira that is uttered in *chrē*. In this version, then, the line reads: 'It is necessary that the utterance and beholding of that which 'is' Is; for being Is'. Beholding that which 'is' is necessary, as we learnt from B8, 34: the *eon* is *noēma* by its nature. *Legein*, as we saw, is the epiphany of the truth of that which 'is' that takes place in the *logos*. The necessity of this epiphany cannot come from anywhere on the outside; it must emerge from *Alētheia* itself. But the origin and meaning of this necessity are still opaque. In order to elucidate them, we will consider the third possible interpretation of the line.
3. The simplest possibility in linguistic terms is to view *to legein te noein te* as the subject and *eon* as a predicate noun. Then the words mean: 'It is necessary that the utterance, like the beholding of that which 'is', Is that which 'is', for being Is.' We already know from B8, 34: it is necessary that beholding Is that which 'is'. For that which 'is' is nothing other than the *noēma*, and *noēma*, as a *nomen rei actae*, is the term for completed and fulfilled beholding. But how should the *eon*, that which 'is', behave towards *legein*? In what sense should one be able to say that the utterance Is that which 'is'? – If we start once again from the linguistic form, it is evident that *legein* and *noein* correspond, that they are in complete equilibrium. It follows from this that the relation of utterance to that which 'is' must not be any different from that of beholding; hence that which 'is' is the fulfilled completion for utterance too. The fulfilled completion of *legein*, however, is the *logos*. This means that the *eon*, that which 'is' as the whole of the well-rounded truth of being, Is the *logos* in which the epiphany of this truth takes place. Thus line B6, 1 confirms the above

interpretation of the words *piston logon ēde noēma* – 'the trustworthy utterance and what is beheld' – from B8, 50.

But this also clarifies the meaning of B8, 35/6:

> For you will not find beholding apart from that which 'is'
> in which it is uttered.

If that which 'is' itself is the *logos* in which the truth of being utters itself, then it is no longer surprising that Parmenides can say that the beholding of this truth is uttered in that which 'is'. And yet these words could lead us dangerously astray. These lines state that what is uttered in the 'word' of that which 'is' is beholding, *to noein*. Has *noein* now taken precedence over *einai*? Has *einai* been subsumed by *noein*? Here too, there is equilibrium, for now the word *noēma* has been replaced by *to eon*. We must therefore cling strictly to the balanced unity of *noein* and *einai*; 'that which is' is, as we are told in B8, 44, *messothen isopales pantē* – 'from the centre, it is equally strong as a whole.'

'From the centre' – *messothen*: what is the centre? Can there still be a centre in sameness? We saw that sameness is sameness in the difference between *noein* and *einai*. What stands as the same in this difference between *noein* and *einai*, stands in the centre, is that which 'is', the *eon* as *noēma*, in which both *noein* and *einai are* within the limits of their fulfilled completion. In this sense, that which 'is' is the 'cohesive' (*syneches*) and the 'common' (*xynon*). But this answer is not sufficient on its own, for the poem speaks here of the centre of that which 'is' itself, that is, neither of the *noein* nor the *einai* nor the *eon* as the *xynon* and *syneches*. The centre of that which 'is': this must be what brings the difference in the sameness of that which 'is' into a state of unity – later philosophy would speak of the 'ground of unity'; in the archaic thinking of Parmenides, however, the 'existent' has not a 'ground' but a 'centre'.

If we are seeking access to the centre of that which 'is', we should start neither from *noein* nor from *einai*; for we are seeking the unifying element that holds *noein* and *einai* together in the *xynon* of that which 'is'. Fragment B5 shows the way:

> But it Is as the common for me and with me:
> whence I begin, thither will I accordingly return.

These words, as we saw, are spoken by *Alētheia*; that which 'is' is the common for *Alētheia* and with *Alētheia*. Because that which 'is' is a *xynon*, the same in difference, its *logos* – which, by revealing that which 'is', Is that which 'is' – has the form of a ring that returns to its origin. We encountered this shape in B8, 34; one could find it running through the entire poem. The element that closes the ring is sameness, in which both *noein* and *einai* are One. But what is this sameness?

According to B8, 35, that which 'is' is *pephatismenon*: it is the 'utterance', i.e. the epiphany of *Alētheia*. It is in *Alētheia*, then, that the *eon*, that which 'is', has its origin as the same in difference. The twofold nature of difference is unified in that which 'is', which is here a same, because its two sides are true in it; beholding is true because being is revealed to it, and being is true because it is revealed to beholding.

For both *noein* and *einai*, being-true is the fulfilled completion through which they are One in the *eon*, which is here a *noēma*. The sameness of the same which both *are* is truth, *Alētheia*. And this is what allows Parmenides to say that the *eon*, that which 'is', is the 'utterance', the epiphany of *Alētheia*. So is *Alētheia* the centre of that which 'is'?

We must not stop at *Alētheia*, however, for B1, 29 states that the poet must experience *Alētheiēs eukykleos atremes ētor* – 'the unshakable heart of well-rounded truth'. If the sameness of the 'existent' is *Alētheia*, then its centre must be the 'heart' of *Alētheia*. But what is the heart of *Alētheia*?

In the light of this question, it is now time to consider that we have so far accepted the most important word in the entire poem as if it were self-explanatory. The most important word in the poem is *estin* – 'it Is'. We encountered infinitive being, *einai*, and saw that it reaches the 'limits' of its fulfilled completion in the participle *eon*, that which 'is'. Line B6, 1 states: 'It is necessary that utterance, like beholding, is that which 'is'; for being Is.' Fränkel's translation makes it clear that the words 'for being Is' are far from self-explanatory; he writes 'for being exists' [*denn ein Sein gibt es*] after stating a page earlier that *estin* here means '(it) is possible (permissible) that...' [*(es) ist möglich (zulässig) dass...*].[21] His translation 'being exists' was discussed above; but the meaning 'for being is possible (permissible)' makes no sense here. For whatever is possible or permissible can also not be; but the words *estin gar einai* – 'for being Is' – explain the preceding *chrē* – 'it is necessary'. Whatever the 'Is' might mean, it is clear that the words say one thing: that the necessity of that which 'is' lies in the 'Is'. In order to elucidate the puzzling meaning of the 'Is', we will compare a different passage in which we encounter the same 'Is'.

The most obvious example of this 'Is' can be found in Fragment B3: *to gar auto noein estin te kai einai* – 'For it Is the same to behold and to be.' – We have so far understood this line as if its two parts were in reverse order: 'to behold and to be Is the same thing', as if *to auto* were a predicate noun. Then the line means that *noein*, having reached fulfilled completion, Is the same, namely the *eon*, that which 'is' as the whole of revealed truth. There is no need to retract any of this explanation. In the meantime, however, we have looked more closely at the meaning of 'the same'; we have discovered that the sameness of this same is the truth that reveals itself in that which 'is' as unity in difference. We have seen that the truth has a heart and that which 'is' has a centre. This centre of that which 'is' must not be understood in Parmenides as the 'ground' of unity; but it is that which, in the balance of forces, makes the same the same and the One the One. Then this centre is 'the same' in a primary sense. Once one has considered the possibility of such a primary same, it soon becomes clear that Fragment B3 also permits, even demands a second interpretation. Within the construction of the sentence, the primary same cannot assume the position of a predicate noun; it must be the subject. But if *to auto* is the subject, *estin* also changes its meaning. Then one has to treat it like a transitive verb: 'For the same Is, i.e. permits to be, both beholding and being.' What is this supposed to mean?

---

[21] Hermann Fränkel, *Dichtung und Philosophie des frühen Griechentums* (Munich: C. H. Beck, 1962), p. 458 and p. 457n9.

As we saw, infinitive beholding and infinitive being 'Is' only by reaching the 'limits' of its fulfilled completion in 'that which is'. It reaches completion through the 'utterance', the epiphany of truth in the *eon*. The *logos*, the 'word' in which this epiphany takes place, is no spoken, sounding word. It is the occurrence of the pure *estin* – 'it Is'. Through the 'Is', the connectedness (the *xynon*) of beholding and being attains the unity of that which 'is', in which truth is then revealed as the sameness of the same. The *estin* – let us repeat it – is epiphany.

But we asked about the heart of truth, the centre of that which 'is', the primary same that can act as a subject for this *estin*. In order to find this primary same, we must once again recall the path that led us from the invocation of the Muses in Homer to this question. The indicator of this path's direction was presentness: as the omnipresence of the divine *nous*, as the omnipresence of being in that which 'is', and as the presentness of this twofold omnipresence in the truth, whose presentness in beholding and being brings both to the fulfilled completion of 'that which is' in the epiphany of the Is. But what does 'present' mean here? The present whose presentness is revealed in the 'Is' is not the same present that appears between past and future in the passing of time and, as soon as it appears, has already vanished again. Nor is it the eternal presence of the Homeric gods, who 'always' are, because they are immortal and remain untouched by change over time. For even this 'always', as we saw from the example of the moon, still relates in its infinite extension to the non-being of the sphere of *doxa*. The only form of present that corresponds to the epiphany of the truth in the 'Is' is the *nyn*, the Now; not the now as a point in passing time, but rather the pure Now of epiphany in which everything is at once and in the One: *oude pot' ēn oud estai epei nyn estin homou pan* – 'and it never was, it never will be; for it Is Now at once and in One as a whole' (B8, 5). 'For being Is' – *estin gar einai*: the combination of infinitive being and the Now results in the 'Is', or, differently put: the 'Is' is being in the Now. In the 'Is', that which 'is' is at once and present in the One Now: being in the Now Is (transitive) that which 'is'. Through the Is, the Now is the enablement of the presence of beholding and the presence of being in that which 'is', which Is Now at once and in one the unity of beholding and being. But the presentness of this presence is truth. Hence the Now is the heart of the truth, whose epiphany takes place in the 'Is'. And therefore the Now is the centre of that which 'is', in which the truth is now present. The 'temporal' meaning of this 'Now' that is not in time, yet of which time is the appearance that becomes visible to us in the rotation of the celestial spheres: this 'temporal' meaning now follows from *homou* – 'at once and in the One'. The Now is neither a point in time nor extension; rather, it Is the unity of time. In this sense, the 'Now', if the word is permitted, is the event [*Ereignis*] of eternity in being. The epiphany of truth that occurs in the 'Is' brings this event to light as its heart. Therefore, the epiphany of truth in the 'Is' is the epiphany of the Now as eternal presence.

Chapter 3
The Irony of Socrates (1971)

> Bear in mind that none of you really knows this man; but now that I have started, I will reveal him to you.[1] What you see is a Socrates who is liable to fall in love with beautiful young men, is always in their company and is greatly taken by them. And then again he is also completely ignorant and knows nothing – so far as outward appearance goes. On the surface you see the moulded form of Silenus. But on the inside, once he has been opened up, you can't imagine, my fellow-drinkers, how much self-control is to be found within. Believe me, he is not a bit interested in whether someone is good-looking, and in fact he despises good looks more than you would ever imagine. The same is true of wealth and every other mark of distinction that most people regard as a matter for congratulation. He considers that all these attributes are worthless and that we ourselves – I mean it – are of no account. He spends his whole life dissimulating (*eironeuomenos*) and teasing people. But when he is in a serious mood and opened up, I don't know if anyone else has seen the statues he has inside, but I saw them once, and they seemed to me so divine and golden, so utterly beautiful and wonderful, that in brief I felt I had to do whatever Socrates told me to do.[2]

With these words, Alcibiades opens the first part of his famous speech in Plato's *Symposium*. As we will see, most later interpretations of the oft-cited but rarely understood 'Socratic irony' up until Kierkegaard use this as their point of reference.

If one were to attempt a thorough interpretation of these lines, one would have to show through an analysis of the *Symposium*'s literary form that they are spoken within a constellation of people, situations and thoughts whose refractions and reflections make the phenomenon of 'irony' appear in numerous aspects. The speech of Alcibiades is a complement to that of Diotima: the seeress from Mantinea had, in an 'ironic' reversal of the typical situation, exposed Socrates's own ignorance and introduced him to the mystery of true eros. But the image of *erōs* that she unveils

[1] This text was first published as: 'Die Ironie des Sokrates'. In: H.W. Wolff (ed.), *Probleme biblischer Theologie: Gerhard von Rad zum 70. Geburtstag*. München: Chr. Kaiser Verlag 1971, pp.383–401. Also in: *Hier und Jetzt* Vol. I, pp. 221–238. The permission to republish this text was granted by Klett-Cotta.

[2] Plato, *The Symposium*, ed. M. C. Howatson and Frisbee C. C. Sheffield, trans. M. C. Howatson (Cambridge University Press, 2008), 216c–217a, pp. 55f.

E. Rudolph and J. Picht (eds.), *Georg Picht: A Pioneer in Philosophy, Politics and the Arts*, Pioneers in Arts, Humanities, Science, Engineering, Practice 19,
https://doi.org/10.1007/978-3-030-31790-4_3

unmistakably bears the features of Socrates.[3] Now Alcibiades paints a picture of Socrates's demonic nature. But what comes to light here is the true nature of the *erōs* depicted by Diotima. The other participants at the symposium (and we, the readers) can discover this, as they have heard Diotima's speech. But it must remain unknown to Alcibiades himself, for he only burst drunkenly and noisily into the banquet after the speech finished, like the satyr chorus after the tragic trilogy (222d). He speaks the truth without realising what he is disclosing. In the lines we are examining, he claims that Socrates despises eros, not realising that the evidence he intends to provide – the story of how Socrates rejected his affection – will show those who listened to Diotima's speech the opposite, namely that Socrates is a true servant of eros.

The opposition between the speeches of Diotima and Alcibiades must, however, be viewed within a larger context, both in literary and conceptual terms. At the start of the banquet (17a ff.), the compulsion to drink was abandoned. Until the forced entry of Alcibiades, who appears in the mask of Dionysus, the companions remain sober (213e). At his behest, the previous rule is overturned. He elevates himself to the role of symposiarch, and now a great feast begins (213e). Thus the work is divided into a sober and a drunken part. Only Socrates appears as the rapturous one among the sober (175c-e), yet is always sober among the drunk, no matter how much he drinks. At the end of the banquet, as clear and alert as always, he leaves his intoxicated fellow revellers, who have been overcome by sleep, and steps out into the light of the rising sun to spend the day in his usual way (223d). In him – though we cannot go into detail here – the *Apollonian* form of *enthousiasmos* takes form. This presents the Dionysian *enthousiasmos* that takes hold of Alcibiades in an 'ironic' light.

But Alcibiades himself also chooses a consciously 'ironic' literary form for his panegyric to Socrates. One of the sophisticated parlour games customarily played at Attic symposia involved each person in succession characterising another with a humorous comparison; the resulting comic speeches were expected to be rhetorical showpieces. This game was known as *anteikazein.* A recollection of comedy shows what literary possibilities such a parlour game held. Alcibiades chooses this form (215a, see 214e) in order to unmask Socrates, as it were. But the very first comparison he makes (215a/b) – the comparison to which the passage quoted above alludes – already shows that even the method of unmasking, when applied to the 'ironic' existence of Socrates, is turned into its exact opposite. For Alcibiades compares Socrates to the statues of Silenus that can be opened up to reveal magnificent figures of the gods inside. With provocative directness, this image shows us the *enthousiasmos* of Socrates. Other people pretend to be better than they are, and appear pitiful when their mask is torn off. Socrates, on the other hand, adopts the guise of a Silenus; yet once he is unmasked, such splendour comes to light that one can only worship him like the manifestation of a god. This paradoxical form of dissimulation is referred to in our passage (216e) as 'irony'. In Alcibiades's interpretation, the irony of Socrates consists in the fact that he poses as one who is constantly beside himself, consumed by erotic desire for beautiful young men. Yet once he is unmasked, it transpires that he disdains not only beauty but all other precious things in life, such as wealth and

[3] 203c/d, p. 40.

honour. His true interior is filled with a *sōphrosynē* ('wisdom'), that rises above all human passions and desires with immaculate radiance.

This concept of irony is far removed from everything we think of when we use the word today. Nor does it align with common notions of 'Socratic irony'. Now, it is a matter of no small importance whether we grasp the basic stance of the man who determined the fate of all subsequent philosophy. First of all, we must clarify what the Greeks themselves meant by the word *eirōneia*. Then we shall examine how the modern concept of irony, which arose during Romanticism, relates to *eirōneia*, in order to free ourselves from assumptions that have placed the image of Socrates in an iridescent twilight in the modern consciousness. Finally, we will prove the thesis that the so-called 'irony' is a reversal of the attitude that the Greeks called *eirōneia*.

I

Aristotle devoted a chapter of his *Nicomachean Ethics* to *eirōneia*.[4] There, irony is described as one form of dissimulation; the opposite form is boastfulness or arrogance *(alazoneia)*. According to Aristotle, a braggart – or a 'show-off', in common parlance – is 'the sort to lay claim to esteemed qualities that he either does not have or has to a lesser degree than he claims. The self-deprecating [ironic] person, on the other hand, seems to disclaim those he has or to play them down. The person at the mean, however, is straightforward, and truthful in life and in what he says, since he acknowledges no more and no less than the qualities he has' (1127a 21ff.).[5] Here we encounter the concept of irony in an unfamiliar context. Aristotle introduces it with these words: '[...] let us now speak of those who pursue truth and falsity in what they say and do – in other words, in what they are claiming' (1127a 19f.).[6] In this chapter, then, he discusses the different forms and possibilities of truthfulness and dissimulation. one can dissimulate by making oneself out to be more than one is; one can also dissimulate by making oneself out to be less than one is. To Aristotle, both forms of dissimulation merit rebuke, whereas 'the truth is noble and praiseworthy' (1127a 29f.).[7] Thus the basic principle of Aristotle's ethics, namely that *aretē* always emerges in the middle between an excess (*hyperbolē*) and a deficient mode (*elleipsis*), is confirmed also in the area of these behaviours.

Before we attempt a closer definition of the term 'irony' in this context, we must rule out all misunderstandings by presenting clearly the differences between Greek aristocratic ethics and Christian ethics. The ethics of the Christian tradition, for reasons found not only in the gospel, but also in the social history of Christianity, projected humility before God into a morality of modesty towards humans. Modesty is demanded without consideration for a person's value or lack thereof; a person of great value is doubly praiseworthy if they also show humility towards their fellow humans. By contrast, it is always considered unvirtuous to entertain a feeling of

---

[4] *Nicomachean Ethics*, IV, 13, 1127a 13ff.

[5] Aristotle, *Nicomachean Ethics*, ed. and trans. Roger Crisp (Cambridge University Press, 2000), p. 76.

[6] Ibid.

[7] Ibid.

superiority. This is known as pride, regardless of whether the sense of superiority is justified or not. The aristocratic virtue of *megalopsychia* – magnanimity – whose description forms the climax of Aristotle's ethics, was viewed with suspicion in the Christian era, at least in the phases defined by the bourgeoisie. It was only in the chivalrous culture of the Middle Ages that it experienced a renaissance. The moral system of 'modesty' gave rise to the suspicion that Christianity was the religion of the disadvantaged, the despised, the powerless, the so-called 'quiet in the land', and simply mirrored their resentment. But this overlooks a second important difference between the Christian tradition of ethics and Aristotelian ethics. Christian ethics in our tradition arises from the question of the soul's certainty of salvation. It resides in the conscience and springs from a self-examination that attempts to illuminate the soul's relationship with God, and is therefore focused mostly on the *inner* comportment of humans in their attitude to themselves and their fellow humans. The fundamental motion of the Christian ethics of conscience: the turn of the soul towards its own inwardness is motivated not only, and not primarily, by particular social situations; rather, it stems from a view of the human being's standing in the face of God that was expressed as much by princes and kings as by servants and beggars. Before God, even *megalopsychia* means nothing, and the awareness of this meaninglessness must be expressed in people's demeanour. Here the entire domain of human behaviour is moved into dimensions so fundamentally different from the horizon of Aristotelian ethics that it can only lead us astray to speak of 'ethics' in both cases.

As the passage on *eirōneia* in particular shows, Aristotle does not view humans from the perspective of inward-directed reflection; he views them as they appear outwardly. This can be seen in the significance of the word that, because our language lacks not only the word but also the way of thinking, I am forced to translate as 'truthfulness' [*Wahrhaftigkeit*]. The Greek counterpart is *alētheuein*. In our context, that means representing oneself as one truly is. The question of whether one is truthful or deceitful towards oneself does not come into the matter; inwardness is not considered. The only concern is whether a person's demeanour, outward behaviour and words correspond to their true nature so precisely that this nature comes forth in its true form. The opposite of truthfulness in the Christian sense is hypocrisy; the opposite of *alētheuein* is dissimulation. But all dissimulation is a manifestation of some flaw.

The virtue that Aristotle is seeking to define is not called 'truthfulness'; it is called *alētheia*). This is the virtue of a man who, as Aristotle puts it (1127b 2), 'is truthful in what he says and the way he lives because that is what his state of character is like.'[8] So it is here that the word *alētheia* has its original meaning, the meaning rediscovered by Heidegger and established against all objections by Wilhelm Luther[9]: it is the 'unconcealment' of the ethos that comes to light in a person's behaviour. Now we must view the form of dissimulation that Aristotle calls

[8] Ibid., p. 77.

[9] Wilhelm Luther, *Wahrheit, Licht und Erkenntnis in der griechischen Philosophie bis Demokrit* (Bonn: Bouvier, 1966).

eirōneia against this background. What matters is not whether the dissimulation is questionable from the perspective of the subject; it is only important that an attitude is adopted which conceals the true nature of a person from the eyes of the world. In this sense, as we have seen, those who pretend to be less than they really are must be called 'ironic'. Such behaviour can, as Aristotle shows (1127b, 22ff.), have its attractive aspects. While boastfulness is unpleasant to others (1127b 8), those who present themselves as worth less than they are convey the impression that their behaviour is motivated not by gain, but rather a wish to avoid pompousness. The English call this 'understatement' [Eng.] – it is the modern equivalent of *eirōneia.* One is so afraid of appearing with bombast and false pathos that one conceals the seriousness, the pride, the gravity and the ambition of one's own standards and pretends that it is all entirely commonplace. While understatement is considered a virtue in our society, however, the Greeks saw such behaviour as such a strong expression of dissimulation and untruth that Theophrastes, in his *Characters*, was able to conclude the chapter on irony with the following words: 'One should be more wary of disingenuous and designing characters than of vipers' (I, 7).[10] Aristotle does say that 'people seem attractive who are moderate in their self-depreciation [irony] and play down qualities that are not too everyday and obvious'[11]; but this statement only refers to the urbane surface, not to what will emerge later as the dangerous and uncanny nature of irony.

But what does all this have to do with Socrates, what does it have to do with 'Socratic irony'? First of all, let us note that Alcibiades uses the word with exactly the same meaning as Aristotle (*Symposium*, 216e 4). He accuses Socrates of dissimulating and concealing, beneath the base mask of a Silenus, the magnificence of his *sōphrosynē*-filled inner self. This is the 'downward dissimulation' described by Aristotle and Theophrastus, and, very much in agreement with Aristotle, Alcibiades rebukes Socrates because he considers his art of dissimulation a way of playing games with people. This is also how Plato uses the word elsewhere; we shall choose the most famous example. In Book I of the *Republic*, Socrates is attacked by Thrasymachus, who accuses him of asking questions only to rebut the answers, while never giving an answer himself. Socrates responds that he is simply unable to find an answer, and should therefore be pitied rather than berated. 'This brought an unpleasant laugh from Thrasymachus. 'Oh my god,' he said, 'I knew it. The irony of Socrates'' (337a).[12] So, just like Alcibiades, Thrasymachus too considers it mere dissimulation when Socrates claims not to know the answers to the questions he asks the others. What we refer to today, in a deeper sense that only arose in the Modern Age, as 'Socratic irony' is precisely *not* understood by Thrasymachus. He too takes Socrates's 'irony' for what the Greeks called *eirōneia*, a mere art of 'dissimulating downwards'. He misreads Socratic ignorance as a mask to conceal knowledge, and this dissimulation, as one repeatedly witnesses in Plato's Socratic dialogues, provokes anger.

---

[10] Theophrastus, *Characters*, ed. and trans. James Diggle (Cambridge University Press, 2004), p. 67.

[11] Aristotle, *Nicomachean Ethics*, p. 76.

[12] Plato, *The Republic*, ed. G. R. F. Ferrari, trans. Tom Griffith (Cambridge University Press, 2000), p. 14.

But what about the true irony of Socrates? Is it merely a modern invention? Or does it have something to do with the 'demonic' (*Symposium* 202d/e) nature of this mediator between human delusion and divine truth (204b)? In the chapter from the *Nichomachean Ethics* we examined, there is a remark that is surprising and difficult to explain in the context (1127b 25): 'And it is especially qualities held in esteem that they disclaim, as Socrates used to do.'[13] Here too, the interpretation follows from the lines from Alicibiades's speech. Alcibiades claims that Socrates only pretends to be a lover of beautiful young men; that, in truth, he despises beauty as much as wealth and honour; he believes that these properties are worthless and that we, who strive for them, are nothing (*Symposium* 216d/c). Beauty, wealth and fame form the heart of the catalogue of highest goods in life that was one of the favourite themes in the archaic poetry of the seventh and sixth centuries, with a multitude of variations, and represents the aristocratic way of life that, even in the democratic Athens of Socrates's day, had still not lost its normative power. From Xenophanes onwards, philosophy began to question the archaic canon of life goods and thus challenged popular opinion. The accusation levelled at Socrates during his trial that he was corrupting the youth can only be grasped against this background; thus the charge that Socrates despised beauty, wealth and honour had a malicious core. One can also understand how the concept of irony could be combined with this accusation. The *alazōn*, the 'boastful person', dissimulates in order to acquire what is valued highly in his surroundings. He pretends to be rich, high-born, educated or whatever else might improve his standing; thus he affirms the values that hold sway. The ironist does the opposite: he acts as if he does not possess the merits that are recognized by society, even if he actually does. He disdains the advantages of having such qualities ascribed to him. How should one explain the fact that he squanders his own benefit? It only makes sense if he scorns all advantage, if he places no value on the love of beautiful young men, wealth or fame, if he believes himself far above all those goods. Thus Aristotle says in the *Rhetoric* that irony is a contemptible attitude because is ridicules what others take seriously (1379b 31; compare 1419b 7). One who seeks honour becomes, through his zeal, an object of mockery if he encounters a man who communicates that he considers honour worthless. The presumed knowledge of the sophists becomes laughable if Socrates presents himself to them as unknowing. This 'downward dissimulation' has the power to unmask 'upward dissimulation'. This irony is especially provocative if the one who dissimulates simultaneously conveys *that* they are dissimulating, if the art of 'downward dissimulation' is practised as a subtle game. It is an expression of the greatest sovereignty if a man can afford to deal in such a way with values that are so esteemed in society, if he can dare to pretend he is unworthy of these values, and when he simultaneously communicates with his demeanour that he considers these values, along with the society that serves them, worthless and contemptible. This is how Alcibiades reads and describes Socrates's behaviour. But even this form of irony is not the true 'irony of Socrates', for Alcibiades, like Thrasymachus, is among those who have *not* understood Socrates.

[13] Aristotle, *Nicomachean Ethics*, p. 77.

The true irony of Socrates only comes to light when one recognizes that Socrates is always telling the truth when others think he is ironically dissimulating. He is telling the truth, his deepest truth, when he speaks of his ignorance; but the others see it as dissimulation, as *eirōneia* in the Greek sense of the word. In reality, he is shaken by the beauty of young men; but eros has taken hold of him in the form revealed in Diotima's speech. That is why his eros appears to be ironic dissimulation, despite being his innermost and tragic drive. Because he has recognized the true use of property and the true radiance of fame, he disdains false wealth and false honour. He is accused of dissimulation because, in a decaying state order, no one can see what Socrates has before his eyes, and the only thing that could save the state. With Socrates, then, irony is not the dissimulation presumed by others; it is the necessary and adequate form in which to articulate the truth he recognizes. It is the expression of a simple ethos, not an insidious, dissimulating one; according to the schema of virtues in the *Nichomachean Ethics*, it belongs not to *eirōneia* but to *alētheia*. Nonetheless, in the state and society in which he is located, his stance must *seem* like dissimulation; it is in the fact that the truth appears to be dissimulation that the tragic nature of this 'irony' lies. It is the precise inversion of the form of speech we term 'tragic irony'. In tragic irony, the speaker themselves is trapped in the deception, but their words have a double meaning and, without realizing, they utter a truth that is understood by the listeners. The speaker operates within a state of delusion, but the listeners know the truth. With Socrates, the reverse is the case: he speaks the truth, but the listeners only hear deception. It has remained thus to the present day: the industrial society of the twentieth century is no wiser than the Athenians who executed Socrates. The tragedy has not gone away, which is why, in every period, the relationship between clear-sighted wisdom and the world surrounding it has remained 'ironic'.

II

The so-called irony of Socrates, then, is neither a particular mode of speech nor the conspicuous character trait of an eccentric Athenian citizen. Rather, it manifests the unbridgeable distance between the recognition of truth at all times, in every society, and the public consciousness with its prejudices. This distance is part of the 'human condition'; hence the universal meaning of Socratic irony. Nonetheless, it is a Greek phenomenon whose true guise can only be ascertained within the horizon of the philosophy of Socrates and Plato. But through the German Romantic school, through Friedrich Schlegel, through Solger and through Kierkegaard's dissertation on the concept of irony, a new and specifically modern concept of irony became fashionable that follows on from Socrates and influenced the interpretation of Socrates among all tendencies influenced by Romanticism, even Hegel's interpretation of Socrates. We must do away with this modern concept of irony if we wish to gain access to the authentic irony of Socrates.

Let us encapsulate the Romantic concept of irony with the words Kierkegaard chooses to explain how Socrates's relationship with the state manifests itself at his trial:

> The objective power of the state, its claims upon the activity of the individual, the laws, the courts – everything loses its absolute validity for him. He divests himself of all of them as imperfect forms; he rises ever more lightly, sees it all disappear beneath him in his ironic bird's-eye view, and he himself hovers over it in ironic contentment, borne up by the absolute self-consistency of infinite negativity. In this way he becomes alien to the whole world to which he belongs (however much he belongs to it in another sense); the contemporary consciousness has no predicate for him – nameless and indefinable, he belongs to another formation. What bears him up is the negativity that still has engendered no positivity. This explains why even life and death lose their absolute validity for him.[14]

We can see at the very first glance that this view of Socrates's irony springs from the same misunderstanding of the Socratic attitude that Alcibiades articulates. The definition of irony as 'infinite negativity' recalls his words: 'He considers that all these attributes are worthless and that we ourselves – I mean it – are of no account.' To Alcibiades, this disregard for all goods and people is a sign that Socrates despises them and elevates himself above them in absolute sovereignty. In the same sense, Kierkegaard speaks of an 'ironic bird's-eye view' and 'ironic contentment'. Similarly, the words 'In this way he becomes alien to the whole world [...] namely and indefinable, he belongs to another formation' seem like a paraphrase of the description Alcibiades attempts to provide of Socrates's *atopia* (215a). Even his words about the statues of the god inside Socrates are mirrored by Kierkegaard directly after the passage above: 'Yet in Socrates we have the actual, not the apparent, pinnacle of irony, because Socrates arrived at the idea of the good, the beautiful and the true only as a boundary – that is, came up to ideal infinity as possibility.'[15]

Comparing the Kierkegaard text to the section from Alcibiades's speech shows that, if I may put it in these terms, the framework of the Romantic concept of irony corresponds to the understanding of Socratic irony espoused by Alcibiades. Yet although the individual definitional elements of this concept are taken from the Greek text, the whole takes on a completely new meaning. We are almost transferred to a different element, and one can define the exact point at which the great reframing takes place. Kierkegaard, like Friedrich Schlegel and Solger before him, draws on the speech of Alcibiades for his definition of irony as infinite negativity. Irony is negative because it rejects goods and people. It is defined as infinite negativity because this rejection expresses the connection of Socrates's existence to the divine images that appear inside him, that is, the idea of the good and hence the absolute. If all seriousness and sincere effort serve only to identify these images of the god, then the existence that stands in such service must distance itself from all other goods and people. It must behave negatively towards them, for all positivity, in this interpretation, has been transferred to the area characterized by Alcibiades's speech as the 'interior of the soul of Socrates'.

But the concepts of 'negativity' and 'inwardness' took on an entirely new meaning in German philosophy after Kant. For negativity is the element that, in Fichte, distinguishes the ‚I' from the 'non-I'; it is thus the fundamental definition of subjectivity

---

[14] Søren Kierkegaard, *The Concept Of Irony With Continual Reference To Socrates*, ed. and. Trans. Howard V. Hong and Edna H. Hong, p. 196.

[15] Ibid, pp. 196f.

and its absolute freedom from the positivity of nature. If irony as a form now enables the representation of negativity, it is an immediate expression of the absolute freedom of the subject, that is, of absolute subjectivity. Then irony is a mirror of the unlimited sovereignty of the genius that comprehends itself in its own subjectivity, and can elevate itself in free play with objects and people above everything that is, even its own empirical existence. That is why Friedrich Schlegel taught that irony is the form of absolute art, because, in the representation of beauty, it simultaneously brings to light the critical distance from each *individual* beautiful thing, namely the infinite freedom of subjectivity. Hegel, in the introduction to his *Aesthetics*, masterfully characterized this concept of irony with a few observations. After first showing how the artist's absolute freedom rises above everything that has value, dignity and sanctity for other people, he continues:

> Moreover, this virtuosity of an ironical artistic life apprehends itself as a divine creative genius for which anything and everything is only an unsubstantial creature, to which the creator, knowing himself to be disengaged and free from everything, is not bound, because he is just as able to destroy it as to create it. In that case, he who has reached this standpoint of divine genius looks down from his high rank on all other men, for they are pronounced dull and limited, inasmuch as law, morals, etc., still count for them as fixed, essential, and obligatory. So then the individual, who lives in this way as an artist, does give himself relations to others: he lives with friends, mistresses, etc.; but, by his being a genius, this relation to his own specific reality, his particular actions, as well as to what is absolute and universal, is at the same time null; his attitude to it all is ironical.[16]

If one looks closely, one can recognise in these lines – which are not without a certain malice towards Friedrich Schlegel – all the traits used by Alcibiades to characterise the irony of Socrates. To be sure, Hegel draws on the concept of absolute negativity and uses the transference of the Greek idea to the medium of modern subjectivity to expand the concept of irony in incredible fashion; it becomes ambiguous, iridescent, indeterminate, seductive, intangible and dangerous. But one need only place our passage from the speech of Alcibiades beside it to see precisely what is happening here, and what mechanism is seeking to cast the spell of Romantic irony on us. Alcibiades, as we saw, already misrepresented the nature of Socrates by describing his behaviour as *eirōneia*, 'downward dissimulation', and interpreting it as an expression of sovereign contempt for all goods and people. The Romantic concept of irony virtually elevates this misunderstanding to a principle, and from this it derives the divine wisdom that Alcibiades was already incapable of grasping. Here, then, the same distortion and falsification of Socrates's image that led to his trial in Athens are heightened. The Romantic concept of irony is the total negation of Socrates. As the negation of Socrates has repeatedly forced its way into the interpretation of his ideas, however, and has also exerted a strong influence on the interpretation of Plato's dialogues, a critical destruction of the image of Socrates it established cannot be avoided on the path we have taken.

---

[16] G. W. F. Hegel, *Aesthetics: Lectures on Fine Art*, trans. T. M. Knox (Oxford & New York: Oxford University Press, 1998–99), vol. 1, p. 66.

III

In the chapter on *eirōneia*, Aristotle says, 'Falsehood is in itself bad and blameworthy, while the truth is noble and praiseworthy' (1127a 29f.).[17] That is why he views *eirōneia* as a vice, even in its attractive forms. In Socrates, however, that striving for truth known since Plato as *philosophia* takes human form. It is inconceivable that the Socratic ethos in particular should be characterised by an attitude viewed by the Greeks as *pseudos*, as deception and dissimulation. If the virtue that Aristotle calls *alētheia* manifests itself in anyone, it is surely Socrates. This is actually confirmed in Alcibiades's speech, for everything he says about the astonishing behaviour of Socrates makes it clear that Socrates was always in agreement with himself and presented himself exactly as he was. This is reflected in the sobriety that, as we saw, is one of the 'leitmotifs' of the *Symposium*, and is underlined by Alcibiades in his speech: 'no living person has ever seen Socrates drunk' (220a).[18] In the earliest panegyric to Socrates in Plato's works, the veteran commander Laches says:

> 'Whenever I hear a man discussing virtue or some kind of wisdom, then, if he really is a man and worthy of the words he utters, I am completely delighted to see the appropriateness and harmony existing between the speaker and his words. And such a man seems to me to be genuinely musical, producing the most beautiful harmony, not on the lyre or some other pleasurable instrument, but actually rendering his own life harmonious by fitting his deeds to his words in a truly Dorian mode, not in the Ionian, nor even, I think, in the Phrygian or Lydian, but in the only harmony that is genuinely Greek. [...] Now I have no acquaintance with the words of Socrates, but before now, I believe, I have had experience of his deeds, and there I found him a person privileged to speak fair words and to indulge in every kind of frankness.'[19]

This is a description of the plain, simple ethos that Aristotle defines as *alētheia* in contrast to *eirōneia*.

Now, the speech of Alcibiades contains a passage in which Plato (with a form of indirect self-quotation that frequently appears in his dialogues) recalls the panegyric of Laches and describes the scene of the retreat from Delion, which had given Laches the possibility to test Socrates. We are told explicitly that Socrates had Laches by his side then, even though there is no other reason to mention the name of Laches in the account. It is with this scene, which constituted the great example of the perfect correspondence between the words and deeds of Socrates, that Alcibiades concludes his report on this man's astonishing behaviour. It forms the capstone and the culmination, yet also contracts his words about *eirōneia*. Socrates did not dissimulate; rather, his mere presence was sufficient to make it clear that almost all other people dissimulate not only to their surroundings, but also to themselves, and pretend to be something other than what they are.

But how does this phenomenon come about that we call the 'irony of Socrates', and which manifests itself in such manifold and subtle ways in the dialogues of

---

[17] Aristotle, *Nicomachean Ethics*, p. 76.

[18] Plato, *The Symposium*, p. 59.

[19] Plato, *Laches and Charmides*, trans. Rosamund Kent Sprague (Indianapolis: Hackett, 1992), 188c-e, pp. 21f.

Plato? What is the true nature of the ethos that Alcibiades misinterprets as *eirōneia*? In Book VI of Plato's *Republic*, there is a description of the philosophers' *physis* (485a–487a) that – how could it be otherwise – is simultaneously something like a portrait of Socrates. In this section, all virtues of the philosophers are traced back to an attitude that Plato calls *apseudeia*: freedom from deception. It consists in the philosophers 'not willingly accepting falsehood in any form: a hatred of falsehood and a love of truth' (485c).[20] This confirms that a truthful philosopher cannot adopt the attitude the Greeks call *eirōneia*; he cannot dissimulate. But now, in the same section, Plato ascribes to the truthful philosopher all those qualities from which Alcibiades had concluded that Socrates is a dissimulator. He writes:

> 'the stronger a person's desires are in one direction, the weaker they will be in other directions. Like a stream when it is diverted.'
>
> 'True. What of it?'
>
> 'In someone whose stream flows in the direction of learning and everything like it, I imagine the desires will be concerned with the pleasure of the mind alone, just by itself. They will give up the pleasures arising out of the body. That's assuming the person is a true philosopher, a genuine lover of wisdom, not a pretend lover.'[21]

The stream symbolises *erōs*. It is this *erōs*, which is gathered in all the philosopher's powers and achieves the ultimate breakthrough in his soul, that explains the erotic behaviour of Socrates which Alcibiades finds so offensive. It explains the distinction between outside and inside, between the body and the soul as it is for itself, which Alcibiades sees as dissimulation. Later on in the same section, Plato writes:

> 'Do you think, then, that the mind which is not afraid of great things, and can contemplate the whole of time and the whole of reality, is likely to regard human life as of any great importance?'
>
> 'No, that's impossible.'[22]

He argues from this that the philosopher places little value on the goods of human life – the second of Alcibiades's accusations. We see that the description of the true philosophers' *physis* returns to the attributes of Socrates we know from the words of Alcibiades. Unlike in that speech, however, they are not interpreted as *eirōneia* but rather attributed to the all-consuming urge towards truth and the incapacity for deception and dissimulation. The point of misunderstanding – if one can use such a superficial word to describe the demonic event of the encounter between Socrates and Alcibiades – lies in the third of Alcibiades's accusations, which we have not yet discussed here, namely that the ignorance of Socrates is mere dissimulation. The mystery of Socrates's irony is identical to the mystery of Socratic ignorance.

[20] Plato, *The Republic*, p. 187.

[21] Plato, *The Republic*, 485d, p. 188.

[22] Ibid.

The statement of ignorance[23] appears in all manner of formulations in the Socratic dialogues, but its most concise form is this: *synoida emautō ouden epistamenō*, 'I am aware that I know nothing') (*Apology*, 22c). The word *synoida* is a technical term from Attic legal language; it refers to the cognisance of an eyewitness. Thus the sentence means: in the trial, which here bears the name *elenxis*, I testify against myself as a witness to the fact that I know nothing. If one looks at the words more closely, one sees that knowledge here appears in two forms: the knowledge of the eyewitness (*synoida*) and the presumed knowledge of the accused who is found guilty through him (*ouden epistamenō*). In this *elenxis*, the interlocutor is not only proved guilty of not knowing; far more importantly, he is forced to discover the first knowledge within himself. What we think we know, but do not, includes not only the content of our supposed knowledge; we also do not know what knowledge itself is, and because we do not know that, we do not know what the *aretē* of humans is meant to be; the norms that govern states and individuals do not withstand *elenxis*. Until this point, the teachings of Socrates correspond to the calm scepticism already espoused by Protagoras; but Socrates radicalises the scepticism of the sophists. But precisely this scepticism led to the great new discovery that ensured the standing of Socrates in the history of philosophy, for he discovered that first knowledge which we encounter as the cognisance of the eyewitness. He discovers that the sceptical unmasking of presumed knowledge presupposes the recognition of a critical standard against which the sceptic can measure the presumed knowledge and expose its hollowness. This fundamental knowledge, hidden deep in the human soul, has a different form from the knowledge that is exposed through the assessment. What is known in this knowledge?

In the *Charmides*, this question – initially still in an aporetic guise – is pursued in the form of an interpretation of the Delphic saying *gnōthi seauton*, 'know thyself'). We know what this saying meant. It means: 'Recognise in the face of the god that you are a mere human.' This insight, which is where all *elegxis* ultimately leads, is referred to in the *Apology* as *anthrōpinē sophia*, 'human wisdom' (20d). It is the awareness adequate to humans of the finitude of their mortal nature. What we must know in our capacity as eyewitnesses in order to expose the finitude and invalidity of our presumed knowledge, then, is the presence before which the finite reveals itself as merely finite. This awareness of boundaries that apply to humans is the subject of the ever-misunderstood Socratic teaching that *aretē* is an awareness. This is not 'ethical intellectualism', but rather a recognition of the invalidity of humans before that god who in Greek philosophy, long before Socrates, had already experienced his great epiphany in the form of the One and Only God (see Chap. 2, p. 17 f).[24]

In Diotima's speech, Socrates himself is now subjected to an *elenxis*. Here, unlike in the Socratic dialogues, it is not the presumed knowledge of the sophists, the politicians, the poets or the Athenian citizens that is exposed in its nullity; rather, it is the knowledge of Socrates, the knowledge of ignorance, that is put to the test. It

---

[23] See Georg Picht, 'Wissen des Nichtwissens und Anamnesis', in *Wahrheit, Vernunft, Verantwortung* (Stuttgart: Klett-Cotta, 1969), pp. 87ff.

[24] See Georg Picht, 'Der Gott der Philosophen' and 'Die Epiphanie der Ewigen Gegenwart', in *Wahrheit, Vernunft, Verantwortung*, pp. 229ff. and 36ff., the latter also in this volume, pp. 7–42.

is a matter of illuminating the form and the inner possibility of *synoida*. By bringing Diotima into play, Plato – in the account of Socrates – gains, by literary means, the possibility of contemplating the figure of Socrates not from the Socratic perspective, but from without and above; Socrates himself is shown in an 'ironic' light. I will not attempt here to solve the mystery of Diotima (on the search for the background to this figure, the academy's cult of muses recalls the muses; her history includes the goddess Aletheia in the poem by Parmenides, as well as the priestesses whom Plato invokes in the *Meno* in his introduction to the doctrine of anamnesis (in 81a); the story of her aftereffects leads, via the personification of Philosophia in Boethius's *Consolation*, to Dante's Beatrice). It is important to note, however, that the appearance of Diotima vividly represents a stage in Plato's own development that allows him to go beyond Socrates, and thus to see the true face of Socrates for the first time. That is why the *Symposium*, both in literary and philosophical terms, is the apex of Plato's depiction of Socrates.

The crucial passage for the determination of Socrates's knowledge is this:

> 'No god pursues wisdom or desires to be wise because gods are wise already, and no one who is wise already pursues wisdom. But neither do ignorant people pursue wisdom or desire to be wise, for the problem of ignorance is this, that someone who is neither fine and good nor wise is still quite satisfied with himself. No one desires what he does not think he lacks.'[25]

This leads to the following definition of philosophy: 'a philosopher is in a middle state between a wise man and an ignorant one' (204b).[26] This very simple statement contains a doctrine that distinguishes Plato's philosophy from all later metaphysics. Metaphysics comes about when knowledge of the 'highest things' attempts to give itself the form of science. But the same knowledge whose nullity is exposed by Socrates takes the form of science (*epistēmē*). In the Modern Age, science even became the epitome of the spiritual blindness that believes it knows what it does not. The teaching of Diotima makes clear what was already hinted at in the early dialogue *Charmides*: the knowledge of the ignorance of supposed knowledge cannot take the form of the science of science, *epistēmē epistēmēs*. It is not thinkable as the possession of knowledge. The form of knowledge that is exposed by *synoida* must not be projected back into the *synoida* itself. The knowledge of *synoida* does not have the form of possession but of striving. It is movement through and through, in every aspect. It does not rest on a *fundamentum inconcussum*, an unshakable foundation. It is not grounded but in limbo, in the middle between wisdom and ignorance. The power that keeps it suspended in this balance is eros. It is the power that carries *philosophia* and propels it. Now Diotima, however, defined this middle with a different pair of opposites. Eros, she teaches, stands in the middle between god and what is mortal; it acts as an interpreter and ferryman between gods and humans (202d/e). In this sense, *philosophia* is 'hermeneutics'. Because eros is suspended in between god and the mortal sphere, it is a *Daimon*. In this sense, *philosophia* is 'demonic'.

It will have become clear now why the words and behaviour of Socrates were bound to be taken for *eirōneia* by his interlocutors. Because he sees through their

---

[25] Plato, *The Symposium*, 204a, p. 40.

[26] Ibid.

purported knowledge and their prejudices, he is infinitely superior to them. They have to ascribe some higher knowledge to him, and can only interpret his ignorance as a mask. The insight that *synoida* does not itself have the form of 'knowledge', but is suspended between god and mortality, between wisdom and spiritual blindness, can only remain inaccessible to them. In his Socratic dialogues, however, Plato depicted this state of limbo, which is incomprehensible to mortal thought, in the words and behaviour of Socrates, and the immeasurable effect of these writings led, in a misunderstanding of the authentic meaning of *eironeia*, to the description of the indirect form in which the knowledge of ignorance must impart itself as 'irony'. Correctly understood, this 'ironic' form of language and overall behaviour is the only form in which to convey a 'knowledge' that is unable to appear as *episteme*, and therefore cannot solidify into a directly communicable 'doctrine'. Yet anyone who misreads Socrates's attitude as 'downward dissimulation' will remain trapped in the error of Alcibiades, namely that Socrates is in possession of the truth and could have it at his disposal if he wanted. They are bound, like Kierkegaard, to view Socratic irony as a self-sufficient sovereignty so strong that even the sovereignty of the Attic state is shattered against its rocks. The reader of Diotima's speech knows that Socrates is *not* sovereign in the knowledge of his ignorance, but is rather in need of initiation. They know that the divinity of what Socrates has in mind is revealed precisely in the fact that he has no control over it, but experiences his superiority in a pathos that is given the name 'eros' in the *Symposium.* Already at the outset of the conversation about *erōs* (177d), Socrates had said that he did not claim to know anything other than what is connected to *erōs.* In other words, the only thing that Socrates knows in his ignorance is the powerlessness of all human knowledge before the appearance of divine truth.

In countering the pseudo-knowledge of humans not with a higher knowledge, but with an ignorance that is aware of itself, Socrates turns *eirōneia* on its head, as it were. It no longer expresses an attitude of prideful superiority over objects of contempt, but rather reflects the stance of a man who refrains from all presumption towards people and institutions whose existence rests on arrogated prejudices. Certainly: viewed from the outside, *eirōneia* and its Socratic inversion are difficult to tell apart. Dissimulation and truth both keep a superior distance from the *communis opinio* (common opinion). The masses only perceive the outward behaviour, and cannot understand the reasons for it. It is inevitable that they will view the inversion of *eirōneia* as *eirōneia* in the usual sense. Thus – to repeat – the irony of Socrates is tragic. His truth appears through the lens of prejudices as dissimulation, just as, conversely, 'tragic irony' means that the deluded person utters a truth of which they themselves are unaware. What we call 'tragic' is a fate that inescapably arises when truth is transferred to the sphere of deception or deception to the sphere of truth. Where deception and truth shatter against each other, the result is that unwanted yet inevitable ambiguity of speech and behaviour that is mistaken for *eirōneia.* It is not dissimulation but an expression of truth that every statement made by Socrates is bound to mean something different to his partners and to himself.

Chapter 4
The Concept of *Energeia* in Aristotle (1959)

The word *energeia* is a neologism coined by Aristotle, derived from *ergon*, 'work'. Scholasticism replaced this with *actualitas*; *actus* refers not to the work, however, but the act.[1] Hence this expresses a far-reaching change in our understanding of that of which we say that it 'is'.[2] As a translation of Aristotle's term, Meister Eckhart gifted the German language with a word that emulates the Greek *energeia*; instead of *actualitas*, he used *Wirklichkeit*.[3] We still speak in Aristotle's language today when we use the word *Wirklichkeit*.

Behind the meaning of such fundamental philosophical terms lie circumstances that can scarcely be identified using the methods of philology or linguistic analysis. We cannot speak of 'actuality' without some prior understanding of what we consider to be the 'true', the 'existent', the 'real', the 'objective' – that is, the indisputable tenet whereby it becomes apparent that a concept in which the nature of God has been thought at once refers to the nature of the world and the nature of ourselves, of humans. This prior understanding informs all the words we use; it cannot be seen and identified only in the word for actuality. But words have a history; if one becomes aware of the worldview from which they arose, then one's own prior understanding too becomes not only that of one word, but the horizon in which it places us also becomes transparent. Then the history of the concept serves one's own world orientation. The *actualitas* in which we are situated only actually becomes visible through the examination of such a word. That is why those who are not professionals in the fields of philosophy or philology should still concern themselves with the meaning of the concept that gave our everyday language words like 'reality', 'actuality' and

---

[1] This text was originally published as: 'Der Begriff der Energeia bei Aristoteles'. In: *Hier und Jetzt* Vol. I, pp. 289–308. The permission to republish this text was granted by Klett-Cotta.

[2] This essay and a supplementary text 'Act and Being in Schelling' were given in 1959 and 1961 as presentations to a committee of physicists and theologians at the Protestant Institute for Interdisciplinary Research (FEST).

[3] The noun *Wirklichkeit*, the standard German word for 'reality' (along with *Realität*), is connected to the verb *wirken*, meaning 'to work, have an effect' (trans.).

E. Rudolph and J. Picht (eds.), *Georg Picht: A Pioneer in Philosophy, Politics and the Arts*, Pioneers in Arts, Humanities, Science, Engineering, Practice 19,
https://doi.org/10.1007/978-3-030-31790-4_4

'energy'. The following reflections have been written for those readers with as little academic paraphernalia as possible.

Aristotle himself explains the concept in Book IX, Chap. 8 of the *Metaphysics*:

> *To gar ergon telos, hē d'energeia to ergon, dio kai tounoma energeia legetai kata to ergon, kai synteinai pros tēn entelecheian.* – For the work is the goal, but the actuality is the work, which is why the name 'actuality' is an account with reference to the work, and it is extended to the entelechy.[4]

Here we find another neologism coined by Aristotle: *entelecheia*, derived from *entelechēs*, also his own invention. Attempting to explain this adjective presents some difficulties, however. As far as I can see, the established opinion today is that of Diels, who stated that the adjective *entelechēs* is derived from the adjective *entelēs*, meaning 'complete'; then *entel-echēs* would be synonymous with *to entelēs echōn*, 'having the complete'. And then *entelecheia* would be the state or condition of having the complete.[5] This explanation is thwarted by the fact that Aristotle by no means uses *entelecheia* to refer to the nature of perfection. As an example, let us consider a passage that is at once suitable to give us a factual understanding of the matter. In Book VIII, Chap. 5 of the *Physics*, Aristotle writes (257b8): *Estin d'hē kinesis entelecheia kinētou atelēs,* 'but motion is an incomplete entelechy of the movable'; *atelēs*, 'incomplete', is the opposite of *entelēs*, 'complete'. If the explanation for *entelecheia* were defined as the constitution which *to entelēs echei*, 'which has the complete', then an *entelecheia atelēs* would cancel itself out. But another explanation is also possible. When the Greeks spoke of 'motion', they meant not only spatial motion (*phora*) but also change (*alloiōsis*). If we understand motion in this encompassing sense as motion towards a goal (*telos*) – where the 'goal' can manifest itself not only as a particular place, but also a particular state – then the desired goal must already be present for the movable in a certain sense; otherwise it could not move towards this goal. If the goal were fully present, it would already have been attained and would no longer be a goal; if the goal were completely absent, the motion could not be directed at it, and it would also not be a goal. The goal is a goal if it is present, but present as something that is absent and not yet attained. Something that is in a state of motion towards a goal 'has' a goal when it has not yet reached that goal. But the way in which it 'has' that goal is precisely a motion towards the goal. That is why motion is the having of a goal without reaching the goal: *telecheia atelēs.* But if being-in-motion is the essence of the movable, it carries within itself the having of a goal without reaching the goal (assuming it is movable at all): *to kinēton en heautō to telos echei.* Then *entelecheia* would have to be interpreted as *en-tel-echeia*, as the 'having-in-itself of the goal'.

The movable has its goal within itself as its *potentiality*. This emerges clearly from a passage from Book III of the *Physics* that corresponds precisely to the one quoted above (201a11): *Hē tou dynamei ontos entelecheia, hē toiouton, kinesis estin,* 'the

[4] Aristotle, *The Metaphysics*, 1050a21ff.

[5] Thus Ross on *Metaphysics* 1047a30 in Liddell and Scott, *Greek-English Lexicon.*

entelechy of that which is[6] in potentiality, in so far as it is such, is motion.' That is, motion is the way in which that which is in potentiality already has the goal within itself by virtue of its potentiality. The *dynamei on* – 'that which exists potentially' – in this statement is identical to the *kinēton* – 'the movable' – from the statement above, for the movable is only movable because it exists within the potential to reach its goal.

Let me point out that this examination of *energeia* has raised another concept that is closely connected, and which we also use as a term for 'energy' today: the concept of *dynamis*, which modern physics adopted in the word 'dynamics'. For Aristotle, the connection between *dynamis* and *energeia* is based on a conception of nature which assumes that *physis* is the 'source of motion', that everything which can exist in nature is only grasped in its being if we conceive it in the motion that is particular to it. For Aristotle, the measure of motion is time. The interpretation of being as being-moved interprets the being of that which is within the horizon of a prior understanding of time that is already implicit in the concept of *energeia*, of 'actuality'. That same prior understanding of time is also contained in the concept of *dynamis*. It is not identical to the implicit understanding of time in modern 'energy' and 'dynamics'; modern natural science knows no entelechy. But we should start to think about how the temporal horizon in which we operate relates to that of Aristotle, and accordingly about the relationship between our 'actuality' and Aristotle's *energeia*.

In order to avoid losing our way, it will be useful to recapitulate our steps so far. We started from Aristotle's explanation of the word *energeia* in Book IX of the *Metaphysics*. Because he says there that *energeia* extends in the direction of *entelecheia*, we were forced to explain this Aristotelian neologism too. But we have not reached the end in doing so, for it remains an open question how we are to think the *telos* – the 'goal' – that gives *entelecheia* its name. Aristotle says, 'The work is the goal' (1050a21). So we explain the meaning of *energeia* by attempting to determine the *telos* in *entelecheia* and, to this end, to examine a text in which Aristotle uses a significant example to explain the nature of *entelecheia*.

At the start of Book II of his work *De Anima*, Aristotle develops a general definition of *psychē*. It is misleading, however, to translate this word as 'soul', for *psychē* is not only the soul in humans, but is what makes a living thing – a plant, an animal or a human – alive at all. Aristotle defines this concept of *psychē* as *entelecheia hē prōtē sōmatos physikou organikou* – 'the first entelechy of a natural organic body' (412b5). The definition is explained by the steps Aristotle takes to reach it.

The 'actuality' of what the actual is in all things actual, namely of life, can only emerge from that which attains it in its actuality: being. Aristotle therefore begins this section with a compressed recapitulation of his theory of *ousia*. This term, derived from the participle of the Greek word for 'to be', means 'beingness'.[7] The meaning of this word comes from the science that Aristotle called the 'First Philosophy';

---

[6] For the translation of the ontological terms see Footnote 2 in 'The Epiphany of the Eternal Present', p. 9 (ed.)

[7] The common English translation of *ousia* is 'substance' or 'essence' (trans.).

since the sixteenth century, we have called it 'ontology'.[8] 'Ontology' is the science *hē theorei to on hē on kai ta toutō hyparchonta kath'auto,* 'that investigates that which is, in so far as it *is*, and that which underlies it within itself' (1003a21f.). This definition already contains the implicit decision that explains why Aristotle is not interrogating being, as is generally claimed, but rather the 'beingness' of that which is. His investigation gains its direction by looking ahead to what underlies that which is, in so far as it is a 'that which is', in itself as its origin. This is indicated by the reference to *ta hyparchonta*, the causes; it contains a reference to the *archai* – the origins or, as they would later be called: the 'principles' of 'that which is'.

Aristotle's decision not to frame the central question of philosophy 'infinitely' as a question of being or, in all-encompassing fashion, of *physis*, but rather as the question of *ousia*, has far-reaching consequences. *To on*, 'that which is', is for him always a *tode ti*, a 'this something', something available that is to be determined in one way or another. This happens in the form of the *logos apophantikos*,[9] of 'declaratory statement', which would later be termed 'judgement' in logic. In this statement that which is stands in the position of the grammatical subject. Its determinations are stated through predicates. In this form of statement, that which is is grasped as the *hypokeimenon* (Lat. *subiectum*), the 'underlying', and defined with a term that, as a noun, has the quality of not referring to time. Predicates, on the other hand, as Aristotle notes in *De Interpretatione* (16b6), also signify time. If one combines the timeless noun with a temporal predicate, this results in a statement that is either true or false. Aristotle refers to the basic forms in which such a combination of the timeless noun with the temporal predicate is possible as 'categories'. In these, one can see within which structures temporal and mutable determinations can be combined with an immutable, self-identical existent. Hence they contain the system of coordinates for the forms of movement within which that which is *is*. The consistency and strict self-contained nature of this doctrine rest on the assumption that the structure of the *logos apophantikos* directly exhibits the immanent structure of the *on hē on* – 'the existent, in so far as it is an existent'. The linguistic distinction between subject and predicate corresponds to the distinction found in every existent between that which is immutable and self-identical within it on the one hand and its changing determinations on the other. Logic is immanent in the existent – that is why statements that are congruent with logical forms can be true. Hence the 'First Philosophy' of Aristotle is indeed an 'ontology'; it is a science whose first positings anchor the truth of the *logos* directly in the structure of the existent.

In the *logos apophantikos*, as we have seen, the existent mentioned here stands in the position of the grammatical subject. It is the underlying, in whose direction all

---

[8] See Footnote 2 in 'The Philosophical Concept of Ethics', p. 130 (ed.).

[9] Editor's note: The *logos apophantikos* states that something which is (subject) has a property (predicate). Unlike a request, command or question, it is a declaratory statement that can be true or false. Aristotle considers it a faculty of the human being – as the being that 'has' the *logos* – to show predicates in relation to that which is, not simply to ascribe them. By stating and in stating, the *logos apophantikos* reveals that the 'that which is' whose properties are being stated behaves of its own accord in the way the statement reveals. Thus, in the *logos apophantikos*, the *logos* of that which is is itself expressed in language. This is the basis for its claim to speak truth.

other determinations are stated in the forms of the categories. That is why all other categories refer back to this first category, which Aristotle terms *ousia*, 'beingness'. Aristotle's fundamental decision to make the participial form 'being' the primary one in which being shows itself in nature was grounded, as we can see, on the prior decision that the categorial structure of the *logos apophantikos* must directly exhibit the being of the existent exactly as it is. Because the predicate states time as part of the form of its statement, whereas the noun, which refers to *ousia*, points to something immutable that can underlie temporal determinations because it is 'without time', the question arises as to what the nature of this *ousia* must be if it is to take up temporal determinations without being affected in its being-equal-to-itself. Aristotle answers this question in the summary of his theory of *ousia*, with which he prefaces the definition of *psychē* given above.

He distinguishes three aspects of *ousia*:

1. *Hylē*. The exact translation of this word is *material*, 'wood'. But when we speak today of 'matter', this immediately invokes all the notions that attached themselves to this term in the natural science and philosophy of the nineteenth century. We will therefore avoid the translation and ask: what does *hylē* mean? Aristotle gives a paradoxical determination that logically excludes any definition: *Hē hylē apophasei dēloutai,* 'the *hylē* is revealed through negation' (1058a23). The reason why the *hylē* is elucidated through negation, and only through negation, can be found in a negative determination: the *hylē* is *to aoriston*, 'the indeterminate'. It is not indeterminate because we have not yet found its determination; rather, it is indeterminate by nature. The *aoriston* – as the name tells us – cannot be revealed in a *horismos*, a definition; one cannot show what it is. Thus the *hylē*, as Aristotle says, is *agnōstos kath' autēn*, 'not per se an object of cognition' (1036a9). Negation is the only logical form in which the unknowable can be identified.

   The ontological character of the *hylē* is determined even more clearly by means of a concept connected to that of the *aoriston*: the *hylē* is the *apeiron*, that 'which has no limits'. Aristotle also interprets this concept, whose history can be traced back to Anaximander and Pythagoras, in the context of his theory of motion. This is the meaning behind his fundamental principle that *to d'apeiron emphainetai prōton en tō synechei,* 'the *apeiron* appears primarily in the continuum'.[10] For according to Aristotle, the continuum is infinitely divisible. We might thus be inclined to say that the *hylē* is the continuum. But this would be a positive determination, and Aristotle therefore states that 'it comes to light in the continuum'. If one seeks to unify the Aristotelian concept of the *hylē* with that of 'matter', one must rid the latter of all connotations that indicate a possible determination of place, time, an attribute or even a relation, for Aristotle explicitly states that no such determination of the *hylē* is possible. Nor can one say, according to Aristotle, that matter is 'in' the continuum of space and time. One could at most say that matter is the continuity of the continuum. But even this, as we saw, would still be too positive a determination. We can only

[10] Aristotle, *Physics*, 200b17/18.

say that the unknowable nature of the *hylē* manifests itself in the continuity of the continuum.

There is a further determination of the *hylē*, however, that takes us to the threshold of the concept of *energeia* and *entelecheia* and would seem to be a positive definition after all. At the beginning of Book II of *De Anima* (412a9), Aristotle says: *esti d'ē ... hylē dynamis,* 'but *hylē* is potentiality'. Looking back at the previous determination, this means that we can also only understand the continuity of the continuum if we grasp the connection between continuity and potentiality. I cannot go into that any further here, however. What is necessary, however, is to point out a logical paradox that is constitute of the concept of *dynamis*. In Book VII of the *Metaphysics*, Aristotle states that everything which comes into being has *hylē, dynaton gar kai einai kai mē einai hekaston autōn, touto d'estin hē en hekastō hyle* – 'it is capable both of being and not being; but this potentiality is in each of them the *hylē*.'[11] Aristotle takes generation as an example to show something that applies to every change, indeed every motion. To explain this, we will divide the process of change into its different phases and refer to the phase in which the mutable is currently located as 'A'. Phase A is then quite simply what this existent here *is* now as something mutable; for the correct answer to the question 'What is that? is the name or definition of A. The being of the mutable is to be in a phase, and at this moment it is either A or not A. This is how Aristotle understands being when, in the course of an analysis of change in Book VI of the *Physics*, he says: *Pan gar anankē ē einai ē mē einai,* 'everything must either be or not be' (235b15). This is the ontological version of the principle of excluded middle. But if we answer the question 'What is that?' with the definition of phrase A, we conceal the fact that the 'this something' currently undergoing this phase is simultaneously a mutable; it is thus determined in its being by the possibility of moving into a different phase. The next phase can be not only phase B, but also the respective multiplicities of phases B, C, D, E etc., each of which is continued in further sequences of possible phases. From phase A, the mutable can only move either to phase B or to phase C, as the principle of excluded middle applies to each particular phase. But the statement quoted above that everything 'is capable both of being and not being' applies to each aspect of being that we had concealed in the determination of the 'is', namely the potentiality contained therein – each of the possible phases (B, C etc.) can either exist or not exist, and precisely this potentiality is its *hylē*, that is, its mutability. The principle of excluded middle does not apply to the *hylē* as such, as the indeterminable surplus that does not exhaust itself in the determination of what 'this something' is.

But this principle can certainly help us to determine the nature of the *hylē* even more precisely. In Book IV of the *Metaphysics* it appears in this version: *Oude metaxy antiphaseōs endechetai einai ouden, all'anankē ē phanai ē apophanai hen kath'enos hotioun* – 'Nor is it possible for there to be something in the middle of a contradiction; rather, we must either assert or deny any single predicate of

[11] Aristotle, *The Metaphysics,* 1032a20ff.

any existent' (1011b23f.). If the *hylē* were an existent, we would have to term it the intermediate excluded in this statement. But because the principle of excluded middle cannot be applied to it, the *hylē* is *aoristos* – indeterminate and indeterminable, and an indeterminable cannot be a 'that which is'. That is why Aristotle states, at the start of Book II of *De Anima*, that the *hylē* is not intrinsically a 'this something'.[12] As the principle of excluded middle refers in Aristotle to that which is, its validity is unaffected by the *hylē*. The *hylē* is never itself a 'that which is; but it is certainly an essential aspect of the 'beingness' of every 'that which is', its *ousia*, whose meaning will emerge even more clearly when we examine the other aspects whose union makes that which is into a 'that which is'.

What does all this have to do with *energeia*? The answer can be found in a definition of the term *dynaton* whereby Aristotle was able to master logically the nature of potentiality, despite its elusion of logic, and construct a theory of inferences of potentiality: *Esti de dynaton touto, hō ean hyparchē hē energeia hou legetai echein tēn dynamin, outhen estai adynaton.* – 'Something is possible if it does not lead to an impossibility if it is given the actuality of that of which it is said to have the potentiality.'[13] We can find a translation of this sentence into the language of modern logic in Albrecht Becker's study on Aristotle's theory of inferences of potentiality: 'An object χ potentially has a property φ if, and only if, the assumption that it actually has this property does not imply anything impossible.'[14]Aristotle does not speak of assumptions *about* the 'object' and its 'properties', however, but argues that the existent itself *is* possible – we would say it 'has' potentiality. And that is why he does not say, 'it has the property', but rather speaks of 'having the actuality of that of which it is said to have the potentiality.' This actuality is its own being-actual in the sense we determined above as the being-in-a-phase of the mutable. Aristotle's definition of the *dynaton*, the potential, has been criticized with the charge that it determines the potential via the negation of the impossible. But this form of definition corresponds to the nature of the *hylē*, which can only be shown through the form of negation.

2. The *eidos*. We now turn to the second aspect of *ousia*, which Aristotle introduces at the start of Book II of *De Anima* with the phrase 'shape and form' [*morphē kai eidos*]. I cannot undertake a full explanation of these two concepts here; it would have to begin with a treatment of Plato's theory of Forms. I will therefore keep to those points that contribute to understanding the concepts of *entelecheia* and *energeia*.

In this passage, Aristotle initially offers only one definition. After saying of the *hylē* that it *kath'auto ouk esti tode ti* – 'is intrinsically not a this something', he now says of shape and form: *kath'ēn ēdē legetai tode ti* – 'on the basis of

[12] Aristotle, *De Anima*, 412a7.

[13] Aristotle, *The Metaphysics*, 1047a24ff.

[14] Albrecht Becker, *Die Aristotelische Theorie der Möglichkeitsschlüsse* (Berlin: Junker und Dünnhaupt, 1933), p. 8.

which a this something is revealed in the *logos* (412a8). The word *legetai* – 'is said' – does not refer here to just any given speech, but to the *logos* that answers the question 'what *is* that?', and which Aristotle therefore also terms the *logos tēs ousias*. In contrast to the *hylē*, which can only be shown via negation, the *eidos* is that part of that which is which can be shown in a *logos*, that is, in a definition. That is why, in Book VII of the *Metaphysics*, Aristotle also refers to the *eidos* as *hē kata ton logon ousia* – the 'beingness' that is appropriate to the logos. The 'this something' is shown in the *logos* because it only gains its determinacy and distinctness from other things through its *eidos* – this matchbox in distinction from this pipe. While the *hylē* manifested itself in the continuum, the *eidos* reveals itself in the structure, which, being determinate, has a *logos*.

Here we encounter a problem, however. Although it is shown in the *logos* what 'this something is', the *logos* cannot grasp the existent in its role as 'this something'. Aristotle says, *ho de logos esti tou kath'olou*, 'the *logos* is an account of the universal' (1035b34). 'This something' is a circle, for example. The *logos* that states what 'this something' is, is the definition of the circle. But the definition of the circle is the *logos* of infinitely many possible circles. This circle here does not have a *logos*; to recognize it, I must augment the *logos* with a perception, namely an intellectual or sensual perception (noesis or aesthesis) (1036a6). Perception captures that part of that which is that is not taken up into the *logos tēs ousias*, namely the *hylē*. Because the *eidos* shown in the *logos* is the universal 'form' of that which is that is determined by the *eidos*, but not the unique form of 'this something', Aristotle, in the passage above, combines – as he often does – the term *eidos* with *morphē*, which refers to the *eidos* that manifests itself in the *hylē*.

3. The *synolon*. This brings us to the third determination of *ousia*, which Aristotle only hints at in our passage from Book II of *De Anima* with the phrase *to ek toutōn*, 'what emerges from these'. The Aristotelian term for this definition is *to synolon* – the whole (*holon*) that exists in the connection (*syn-*) of *hylē* and *eidos*. The later translation of *synolon* is *concretum* – that which has grown together. What this means is evident in a passage from Book VII of the *Metaphysics* that encapsulates the entire principle of *ousia*: *Hē ousia gar esti to eidos to enon, ex hou kai tēs hylēs hē synolos legetai ousia*, 'the *ousia* is the indwelling *eidos* from which, together with the *hylē*, the concrete *ousia* is revealed in the *logos*' (1037a29f.). The phrase 'the indwelling *eidos*' testifies to Aristotle's criticism of Plato, who would have said, 'the *ousia* is the *eidos* in itself'. It will thus be important to determine more precisely how this 'indwelling' of the *eidos* in the *hylē*, which turns the *on* into a *synolon*, is to be envisaged. Aristotle encapsulates the relationship of the *eidos* or *morphē* to the *hylē* in a concise formulation in Book II of the *Physics*: *hē physis dittē, hē men hōs hylē, hē d'hōs morphē, telos d'autē* – '*physis* is twofold: on the one hand as *hylē*, on the other hand as *morphē*, but the latter is the *telos*' (199a30f.). The *morphē* – the shape – as we have seen, is the indwelling *eidos*. But the *eidos* cannot be in the *hylē* in such a way that it is entirely subsumed by the *synolon*. For the *eidos*, as we have seen, is the universal; the *eidos* of 'this something' here is identical to the *eidos* of

infinitely many other possible 'this something' of the same kind, for they all have the same *logos tēs ousias*. But if the *eidos* is the shared *telos*, then unity can be understood in diversity.

But does Aristotle, in defining the *eidos* as the *telos*, not abandon the indwelling of the *eidos* once more? For the *telos* necessarily always lies beyond that which strives for it, and it is precisely the beyondness of the *telos* that explains its universality, that is, its identity within infinite diversity. The definition of the *eidos* as the *telos* seems to lead us back to Plato.

It is to this objection that Aristotle responds with the concept of *entelecheia*. In our passage at the start of Book II of *De Anima*, he writes: *esti d'hē men hylē dynamis, to d'eidos entelecheia* – 'but the *hylē* is potentiality, whereas the *eidos* is *entelecheia*' (412a9f.). We see that the concept of *entelecheia* is synonymous with *to eidos to enon* – the indwelling *eidos*. *Entelecheia* is the having-in-itself of the *telos*; the *eidos* is 'in' the *synolon* because the *synolon* has it in itself as the *telos*. The Aristotelian concept of *entelecheia* thus replaces the Platonic concept of *methexis* – partaking of the *eidos* – that Aristotle criticized. He not only replaces *methexis* with *entelecheia*, however, but goes a step further: he says that the *eidos* is *entelecheia*. That is to say, the *eidos* as *telos* does not intrinsically have a being, separate and independent from the existent that strives for it; rather, it can only *be* at all in the *entelecheia*, that is, in the way in which the *synolon* has its *telos*. But in what way does the *synolon* have its *telos*? What is gained through Aristotle's replacement of *methexis* with *entelecheia*? We recall the passage from Book III of the *Physics*: 'the *entelecheia* of that which exists in potentiality, in so far as it is such, is motion' (201a11). Every *synolon* is 'that which is potentially', a *dynamei on*, by virtue of the *hylē* that determines its 'being-this-something'. Hence we can say of every *synolon* that its *entelecheia* *is* motion, that is, the being-moved and being-movable that constitute its nature. Motion in the movable that strives for a *telos* is itself the indwelling of the *eidos enon* in the existent.

Now we are equipped to understand Aristotle's definition of the 'soul' quoted above. The *psychē* is something that manifests itself in connection with a *sōma*, a body. It is not connected to mathematical or artificially produced bodies, however, only a natural body, and among the natural bodies, only those that possess life (*zōon echei*, 412a13). What Aristotle means by 'life' – *zōē* – is nourishment, growth and expiry through itself.[15] 'Through itself' – this means that the origin of the motion that takes up nourishment, that grows and expires again, lies in the living itself. As Aristotle puts it: *en heautou archēn echei kinēseōs kai staseōs* – 'it has in itself the origin of motion and rest' (*Physics*, 192b13f.). But this is precisely what, in an outstanding sense, makes the animate body a natural body, a *physikon sōma*, because for Aristotle, *physis* is the source of motion – *archē kinēseōs*.

The 'that which is' in which we find the soul is the *physikon sōma echon zōēn*, 'the natural body that possesses life'. It is ensouled precisely because it possesses

[15] *di'autou* – that is how 412a14 should be read, as Themistius argues.

life. Taken on its own, it is only a *physikon sōma* ***dynamei*** *zōēn echon*, 'a natural body that potentially has life'. In the 'that which is' that we called the 'natural body that possesses life', then, the mere corporeal body is thus the aspect of potentiality, of *hylē.* Aristotle concludes from this: *anankaion ara tēn psychēn ousian einai hōs eidos sōmatos physikou dynamei zōēn echontos* – 'It is necessary, then, for the soul to be *ousia* in the sense of the *eidos* of a natural body that potentially has life' (*De Anima*, 412a19f.). But how can the soul be the *eidos* of the body? As we saw, the *eidos* is in the *synolon* as *entelecheia*, and *entelecheia* is motion towards the *telos*. The animate natural body is ensouled because it has the source of its motion towards the *telos*, its *entelecheia*, within itself. But we will not content ourselves with this just yet, for we said above that the *eidos* appears in the concrete existent as its shape (*morphē*). So if the *psychē* is the *eidos* of the animate natural body, it must also determine its shape.

Aristotle defines the parts of the shape of an animate natural body as *organa* – as 'instruments'; we still say 'organs' today. The reason for this definition can be found in Book I of *Parts of Animals*: *to men organon pan heneka tou, tōn de tou sōmatos moriōn hekaston heneka tou* – 'every instrument is for a purpose; but of the parts of the body, every one is for a purpose' (645b14ff.). The purpose of an instrument is the work it serves, the *ergon* that gives the *organon* its name; this work determines its establishment and its shape. The parts of the animate body differ from those of an inanimate body in that they are instruments existing for the sake of a purpose, and that they owe their establishment and shape to this function. That is why they are termed *organa.* But what is the purpose? What is the work that emerges? Aristotle continues: *to d'hou heneka praxis tis,* 'but the purpose is a sort of completion [*praxis*].' Aristotle's use of the term *praxis* is determined by its opposition to *poiēsis*, a 'bringing forth'. *Poiēsis* is a making in which something is produced, such as the building of a house. The structure of *poiēsis* is determined by the fact that the goal, the produced work, lies beyond the act of bringing forth as such. The work breaks away from the process of production and now continues its existence as itself; but the act of producing ends when the work is completed. The structure of *praxis* is different; here the consummation itself is the goal. Aristotle explains this with reference to hearing and seeing.[16] Hearing and seeing is not a production that ends when the work is finished, but rather a praxis whose work and goal lie in the praxis itself, namely hearing and seeing. The *organa* of the animate body differ from the majority of other instruments in that they are not instruments of *poiēsis* but of *praxis*. But if the parts of the body serve to bring about a praxis that carries its work within itself, then the same must apply all the more to the whole body. That is why Aristotle, in the passage from *Parts of Animals*, continues: *phaneron hoti kai to synolon sōma synestēke praxeōs tinos heneka plērous* – 'it is evident too that the whole body has been composed for the sake of a fulfilled [*plēres*] *praxis*' (645b15f.). Here, as elsewhere, *plēres* is synonymous with *teleios*; it refers to that which fulfils itself in itself because its goal is itself. The *praxis* of the animate body is fulfilled in this sense because the praxis

[16] Aristotle, *On Sophistical Refutations*, trans. E. S. Forster (London: Heinemann, 1965), 178a9, p. 109.

itself is its work. This praxis, for whose sake the animate body with its parts, the *organa*, is constituted (*synestēke*), is nothing other than its *life*. But the origin of the motion that we can now define as the *praxis* of the animate body is, as we have seen, the soul. It is its *entelecheia*, just as seeing is the *entelecheia* of the eyes.

Ei gar ēn ho ophthalmos zōon, psyche an ēn autou hē opsis – 'for if the eye were a living being, sight would be its soul' (*De Anima*, 412b18f.) This means that the soul of the animate body, as its *entelecheia*, is nothing other than the *praxis* of its very life. From this results the complete definition of the soul: it is *entelecheia hē prōtē sōmatos physikou organikou* – 'the first entelechy of a natural body whose parts are *organa*.'

The Aristotelian definition of *psychē* brought the connection between *praxis* and *entelecheia* to the fore. This connection is made by Aristotle at an important point in his treatise on *energeia*: *ekeinē (hē kinesis) hē enyparchei to telos kai [hē] praxis,* 'the motion that contains its *telos* is also *praxis*' (*The Metaphysics*, 1048b22). The words *enyparchei to telos* are a paraphrase of the concept of *entelecheia*; they confirm that this concept is to be envisaged as *en-tel-echeia*. As examples of such *praxis*, Aristotle introduces seeing (*horan*), sober thinking (*phronein*), the pure observation of truth (*noein*), life (*zēn*) and happiness (*eudaimonein*). He continues: 'such *praxis* is what I call *energeia*.' From this we can derive the following definition: *energeia* is a praxis that contains its own *telos*. It is better, however, if we return after this long digression to the definition of *energeia* that formed our point of departure:

> For the work is the goal, but the actuality is the work, which is why the name 'actuality' is an account with reference to the work, and it is extended to the entelechy.[17]

We now know that what Aristotle is referring to in these lines is *praxis*, the praxis that contains its own *telos*. This *telos* that is contained in the *praxis* is the work it consummates. What Aristotle calls *praxis* consummates itself as its own work. Thus he refers to it as *energeia*, 'being-in-the-work'. *Energeia* is the constant setting-into-work of being-in-the-work. This motion does not end, for it is kept going continuously by the *telos* immanent in it. Hence Aristotle says it is 'extended to having the *telos* within itself'. We can understand these words by recalling that *entelecheia* in the *ousia* of the *synolon* was defined as *to eidos to enon*, 'the indwelling *eidos*'. The *eidos* here determines this something here as what it is precisely because it is not only this something here, but rather points beyond it as the universal that is striven for within it. Hence the 'having' of the *eidos*, namely *entelecheia*, should not be understood as the 'having' of something available, but as the 'having' of a goal towards which the existent must strive as something in motion. And this motion towards the goal of the indwelling *eidos* is 'being-in-the-work', *energeia*. Thus the concept of *energeia* unifies the three aspects of the *ousia* of the concrete, movable existent. *Energeia* is the being of the *ousia* of everything that has the origin of its motion in itself, i.e. everything that is *physei*. In this sense, *energeia* is reality [*Wirklichkeit*] – not as the epitome of everything that is real, but as the nature of the real, as that which allows the real to be real.

---

[17] *The Metaphysics*, 1050a21ff.

In our discussion of the theory of categories above, we encountered the problem of how predicates, which include the expression of time, and hence present that which is as movable, are meant to be connected to a subject that remains self-identical; that is why its description – the noun – is timeless. This problem is solved by the concept of *energeia*, as it uses an idea of motion to characterise the form in which the existent preserves its self-identical nature. This can be demonstrated in the organic body: all movements of its organs that can, in accordance with the categories, be characterized with changing predicates, service the overarching goal of preserving the life of the organism, which means its self-identical nature. Life is the *ousia*, the 'beingness', of this existent; it underlies everything else that can be said about it. One can also see from this example how misleading it is to replace the Aristotelian term *ousia* with *hypostasis*, namely 'substantia'.

But here we encounter a problem that we have so far passed over: the movable cannot be examined in isolation. Precisely because it is movable, it exists in an interaction with other movables, and it is only in this interaction that it *is*. The concept of *energeia* is only suitable to grasp the reality of the real if it also enables the elucidation of this interaction.

So we will not examine the movable now in so far as it has the source of its motion within itself, but in so far it moves or is moved by other things. The difference between these two forms of motion is reflected in Aristotle's distinction between two concepts of *dynamis*. The one we have considered so far refers to the potentiality in the *hylē* to realise the *eidos* that is immanent in it; it refers to the relation of the *hylē* to the *eidos*. In Latin scholastic language it is called *possibilitas*. Aristotle distinguishes another concept of *dynamis*, the one known in scholastic language as *potentia*. This is *archē metabolēs en allo ē hē allō* – 'an origin of change either in something else, or in so far as it is something else.'[18] Aristotle illustrates this with an example: a doctor heals a sick man through his ability to change a state of illness into a state of health. This ability comes from his medical science; he is thus the source of a change in something else, namely the sick man – 'of change in something else'. But he can also heal himself if he is sick. But then he is healed not *qua* a doctor but *qua* a sick man. Then he is a source of change in himself, but while being the other at the same time, namely a sick man – change 'in so far as it is something else'. *Metabolē*, 'change', refers to the transition of something mutable from one phase into another phase; the concept is so broad that it even includes a shift of location from a point A to a point B. In this encompassing sense, *dynamis* is the ability to cause a motion – whether a shift of place or a change – in another. What natural science today calls 'energy' is thus a specific from of *dynamis*.

Dynamis in its manifestation as *potentia* establishes a relation between at least two elements. Wherever there is *dynamis*, there is always (1) the one, namely the moving and (2) the other, namely the moved. In their own ways, both elements are in motion: the moving because it sets things in motion and the moved because it is set in motion. Aristotle calls the motion of the moving *poiein*, meaning 'bringing forth', and the motion of the moved is termed *paschein*, meaning 'suffer' or 'be

[18] Ibid., 1046a10f.

affected'. *Dynamis* as *potentia* is the capacity for *poiein*. On the other side, however, is a corresponding capacity for *paschein*. The moving can only move something that has the potentiality to be moved. Aristotle defines the capacity for *paschein* as *archē metabolēs pathētikēs hyp' allou ē hē allo* – 'the origin of being changed by an affection either through something else or in so far as it is something else' (1046a12f.). For our causal way of thinking, it is remarkable that even the capacity to be moved is defined as a source of motion. Ontologically, however, this definition is irrefutable; for the movement of the body x from point A to point B, its potentiality to be moved is as decisive as the push that moves it from A to B.

Movement as such is only possible if the capacity for *poiein* coincides with a capacity for *paschein*; *poiein* and *paschein* are strictly correlated and mutually dependent. That is why Aristotle says, *phaneron oun hoti esti men hōs mia dynamis tou poiein kai paschein* – 'it is now evident that in one regard, there is a single *dynamis* for *poiein* and *paschein*' (1046a19f.). But does this not lead to terminological confusion? The ability to move and the ability to be moved were defined correlatively in the sense that each required the 'other' for its definition, and it was precisely this relation to the other that constituted the difference between *potentia* and *possibilitas*. But when Aristotle says here that, in a certain sense, the two capacities are a single *dynamis*, he is speaking of *dynamis* as something that is no longer defined in relation to the other; now he means a *dynamis* that is not the *dynamis* of the one part of the correlation or the other, but rather the potentiality of the movement posited with this correlation as such, that is, its *possibilitas*. A similar move from *potentia* to *possibilitas* appears a number of times; hence Bonitz and Ross criticize Aristotle because he distinguished these two meanings of *dynamis*, yet proved unable to formulate that distinction convincingly.

In order to resolve this, let us cross-check this with the corresponding concept of *energeia*, which Aristotle examines in Book III, Chapter 3 of the *Physics*. As the correlation of *poiein* and *paschein* presents two elements, one moving and one moved, we must interrogate both as to their *dynamis* and their *energeia*. Before the onset of motion, the moving is only a moving in terms of its potentiality; it has its *potentia* as *possibilitas*. This *potentia* becomes *energeia* when the moving truly begins to move. Conversely, the moved also has the *potentia* to be moved before motion occurs, namely as its *possibilitas*; its being-moved is its *energeia*. But both are now intertwined, for the motion in which the one moves and the other is moved is one and the same, just as the way from Athens to Thebes is the same as the way from Thebes to Athens. The following lines must be viewed in context:

> *estin hē kinesis en tō kinētō. entelecheia gar esti toutou hypo tou kinētikou. kai hē tou kinētikou de energeia ouk allē estin. dei men gar einai entelecheian amphoin. kinētikon men gar estin tō dynasthai, kinoun de tō energein, all' estin energētikon tou kinētou, hōste homoiōs mia hē amphoin energeia.* – Motion *is* in the movable; for it is its *entelecheia*, under the effect of that which can move it. And the *energeia* of that which can cause motion is no other; for it must be the *entelecheia* of both. That which can cause motion can do so because it has the corresponding *possibilitas* [that is, it has its *potentia* as *possibilitas*]. But it causes

> motion because it is within its *energeia*. However, it can cause motion because it causes the *energeia* of the movable. Therefore, the *energeia* of both is, in the same way, one and the same.[19]

The self-identical *energeia* of the two correlatively intertwined elements is the internally-unified motion in which the one moves and the other is moved. But if *energeia* is one, Aristotle is also justified in saying that, in a certain sense, the twofold *potentia* of the two elements is a single *possibilitas*, namely the possibility of the movement that joins them.

It is hardly surprising that commentators have had difficulties explaining the distinction between *potentia* and *possibilitas* if one considers the boldness of the step taken here by Aristotle. The ontological foundation of the concept of *dynamis* as *possibilitas*, as we have seen, lies in the distinction between *hylē* and *eidos* in one and the same existent. The concept of *dynamis* as *potentia* stems from the fact that every existent has, in nature, an interactive relationship with another existent that it moves and by which it is moved. If I observe the motion of the stone I am throwing, what becomes apparent is (1) the relation between myself and the stone; (2) my ability to throw it (the *potentia* of my *poiein*); (3) the general potentiality in my *hylē*, my body, to move something else (*possibilitas*); (4) the potentiality of the stone to be thrown (the *possibilitas* of its *paschein*); (5) the properties of the stone (such as its weight) determining the trajectory of its flight. But the motion of the stone, although it results from such diverse determinants, is internally unified. It is a single *energeia* of which one can no longer say that it is the *energeia* of the thrower or the stone. It only becomes possible through the relation between thrower and stone, certainly, but neither element in this relationship is its substrate. The ontology anchored in the individual *on he on* thus steps beyond its own boundaries; we are forced to acknowledge an *energeia* that we cannot describe as the *energeia* of a 'this something'. The *ousia* has been replaced by a relation. Does a relation still have a substrate beyond the *relata*? This is precisely what Aristotle claims when he introduces a concept of *possibilitas* that is superordinate to the two linked *potentiae* of *poiein* and *paschein*. *Physis* as a whole, which contains the individual existent present before us as 'this something', is superordinate to it. Thus Aristotle, by introducing the new concept of an overarching *energeia* that can no longer be ascribed to each individual existent, enables the transition to an ontology of *physis* as a whole.

He makes the following remarkable statement about *physis* as a whole: *Hē physis en tautō genei tē dynamei. Archē gar kinētikē, all'ouk en allō all' en autō hē auto* – 'The realm [*genos*] of *physis* is in the same area as *dynamis*, for it is the origin of motion, though not [like *potentia*] in something else, but in the same, in so far as it is the same.'[20] (*Genos* should not be understood here in the sense of 'genus', for there is no overarching genus for the 'species' of *physis* and *dynamis*. The word is interpreted through the phrase *archē kinētikē*. *Genos* is the source from which something originates; for *physis*, this source lies within itself.)

---

[19] Aristotle, *Physics*, 202a13ff.

[20] Aristotle, *The Metaphysics*, 1049b8ff.

According to Aristotle's teachings, the *energeia* whose *dynamis* is *physis* as a whole is the divine *nous*, as the unmoved mover of everything that is in motion. It is thus envisaged as pure *energeia*, as *noēsis noēseōs*. This means that we have reached a threshold where we would have to move from an isolated analysis of basic ontological concepts to a description of the overall conception of Aristotle's physics in its connection to his theology. This would go beyond the scope of this investigation. I will therefore restrict myself to a few remarks that can serve to determine the ontological framework of this conception. The idea that *physis* as a whole should be envisaged as *dynamis*, that is, as the *hylē* whose entelechy is the pure *energeia* of the divine *noēsis noēseōs*, is based on a transfer of the relationship between body and *psychē* to the cosmos. It thus uses Aristotelian terminology to interpret the idea presented by Plato in the *Timaeus* that the cosmos is a divine living being. The methodological upshot of this is that we must search for the key to understanding Aristotelian physics and theology in *De Anima*. *Psychē*, as we saw, means 'life'. When Aristotle says in the passage from the *Metaphysics* quoted above that *physis* is the source of motion 'in the same, in so far as it is the same', he is applying his definition of the living to it: *en heautō archēn echei kinēseōs kai staseōs*, 'it has in itself the origin of motion and rest'.[21] Hence *physis* as a whole, as Plato already taught, has a *psychē*. Only thus can it be *archē kinēseōs*, the 'source of motion'. The reason why Aristotle does not explicitly describe the existing theory of the world soul is that he strictly avoids any hint of mythical language. nonetheless, he presupposes it throughout.

The soul is 'the first entelechy of a natural organic body that has life by virtue of its *dynamis*' (*De Anima*, 412a27f.). In our explanation of this definition above, we neglected to ask why this entelechy is the first; nor did we say what the word *dynamis* means here. To begin with the second point: *dynamis* does not mean that the natural body can be envisaged without life and only with the possibility of having life at some later point. The 'having' of life is rather explained by Aristotle's concept of *hexis*. This is the ontological constitution of the whole body that results in its parts being 'organs'. This ontological constitution is characterized as *dynamis*: life is 'being-in-potentiality'. Hence the words *dynamei zōēn echontos* mean 'which, by virtue of its *possibilitas*, has the ontological constitution we call 'life'.' So why is the 'soul' the '*first* entelechy'?

In Aristotle's definition of motion we have already encountered the concept of *entelecheia atelēs*, the 'incomplete having-in-itself of the goal'. This is the basic form of the mode of being-moved-in-oneself that we call 'life'. In all its manifestations, 'life' is an incomplete having-in-itself of the goal; because it is this, it carries the source of its being-moved in itself. The same applies to *physis* as a whole. What was said about the meaning of the terms *dynamis*, *zōē* and *hexis* in Aristotle's definition of the *psychē* can be applied to *physis*. This explains the statement that *physis* as a whole has its origin in the same realm as *dynamis*.

But Aristotle does not only know the concept of *entelecheia atelēs*, for this can only be envisaged from the perspective of a having-in-itself of the goal that reaches

[21] Aristotle, *Physics*, 192b13f.

and fulfils this goal: entelechy in the primary sense of the world. Here the *telos* is simultaneously the *ergon*. The *entelecheia* is pure *energeia* that has consumed all *dynamis*, as it were, and is thus eternal and immutable. It is into this form of pure *energeia*, the complete and in this sense second entelechy, that the *psychē*, according to Aristotle, transcends in pure *noein*. This can be explained with reference to Book III of *De Anima* and the metaphysical theology from Book XII of the *Metaphysics*. For now, it is sufficient to have shown that the semantic horizon of the ontological concept of *energeia* lies in the region that Aristotle calls *psychē*. That is why he can say: *Hē psyche ta onta pōs esti panta* – 'the soul *is* in a way everything that exists'[22]; he can say *holōs de ho nous ho kath' energeian ta pragmata*, 'the *nous* in the mode of *energeia is* the facts'.[23] We call the congruence of *nous* and facts the truth. At the same time, as we have seen, *energeia* is purely a concept of motion. It forms the culmination of Aristotle's metaphysics because it allows us to envisage the two basic themes of this metaphysics – motion and truth – as one and the same.

[22] Aristotle, *De Anima*, 431b21.

[23] Ibid.

Part II
Ethics, Politics and Right

Chapter 5
Kant's Transcendental Grounding of International Right (1971)

Philosophy has lagged far behind the great questions of our time.[1] Since Nietzsche's collapse, no other philosopher has been able to advance into the force field of the energies that have transformed the face of the earth in the twentieth century.[2] As far as I can see, philosophy no longer ventures onto the terrain of international politics. It has emigrated from politics and surrendered its traditional territory to ideologies. We become aware of this when we try philosophically to grasp the problems affecting international right today. By retreating from the universal questions of metaphysics, philosophy has also lost sight of the horizon of universal history. Until a new form of universal reason emerges, however, there is a fixed relationship between these two manifestations of the universality of thought. That is why the problems of international right have so far only been understood as philosophical problems where the universality of metaphysical thought has come into contact with the universality of the historical horizon. They are located at the intersection of these two hemispheres of the philosophical understanding of the world; yet there are only rarely historical constellations in which the field of view opens up in both directions at once. Kant is the only thinker in whose work the transcendental meaning of international right comes to light, and who was able to determine the systematic locus of its problems. It is worth the effort to consider why this no longer succeeded in later times.

The kairos of Kant's philosophy can be described with the observation that *three* dimensions of universality overlap in his philosophy: the universality of traditional metaphysics, which entered its greatest crisis in his work; the universality of Newton's

---

[1] This text was first published as: "Kants transzendentale Grundlegung des Völkerrechts". In: *Aufrisse. Almanach des Ernst Klett Verlages 1946–1971*, Stuttgart: Ernst Klett Verlag 1971, pp. 223–279. Also in: *Hier und Jetzt* Vol. I, pp. 21–56. The permission to republish this text was granted by Klett-Cotta.

[2] The German word used here, *Völkerrecht* (literally 'law of peoples', 'right of peoples' or 'right of nations'), is normally translated as 'international law'. Because this essay deals with Kantian thought, however, the phrase 'international right' has been used throughout, in keeping with the customary (albeit unidiomatic) translation in this context of *Recht* as 'right' (trans.).

E. Rudolph and J. Picht (eds.), *Georg Picht: A Pioneer in Philosophy, Politics and the Arts*, Pioneers in Arts, Humanities, Science, Engineering, Practice 19,
https://doi.org/10.1007/978-3-030-31790-4_5

celestial mechanics, whose transcendental possibility needed to be substantiated; and the universality of the cosmopolitan ideas of the eighteenth century that emerged from the 'crise de la conscience européenne' (Paul Hazard) in which European thought discovered the insular particularity of 'Christian occidental culture'. In Kant, the crisis of metaphysical thought is inseparably tied to the crisis of the European historical consciousness. The physicist, who had developed the theory of the formation of the solar system and occupied himself with geography and ethnology throughout his life, followed the model of Kepler and Newton in his search for a 'guideline' for understanding history. Reflection on the spherical shape of Earth and the finitude of its surface is fundamental to his idea of cosmopolitan right and the necessity of avoiding wars, as with Herder. That philosophy subsequently lost sight of the breadth of this horizon is connected to the fact that metaphysics, natural science and the philosophy of history dissociated from one another. Thus each of these fields exited the medium of possible universality.

But the restriction of the intellectual horizon of philosophy is also explained by the course of European history; for this led, as Friedrich Meinecke put it, 'from cosmopolitanism to the nation state'. In Fichte and Hegel, *political* philosophy becomes mere *state* philosophy again; Schelling's further development of the Kantian approach went unnoticed. Marx's thought does not live up to its universalist programme; it was largely preoccupied with the economic and social antinomies within European industrialized nations. Nietzsche's vision of the twentieth-century world was not understood. In other ways too, philosophy barely acknowledged the tendencies that determined the development of international politics in the nineteenth and twentieth century: imperialism and colonialism, the expansion of the global economy and its crises, the emergence of new political ideologies with claims to universality, the qualitative change of military power caused by nuclear weapons, the establishment of global communications systems and the resulting modifications of international awareness and the mentalities that inform one another within it. Philosophical thought has been equally oblivious to the fact that the great world religions have entered a new configuration of intellectual powers, since the science of the Modern Age projected its latent ways of thinking onto the screen of an atheist worldview that is now leading a militant life of its own as a political power of the highest order. This applies, at least, to philosophy in the strict sense of the word. There is no shortage of hyphenated philosophies that offer formulas for all things current; but they offer only limited insights, for they avoid the 'work of the concept'. Marxism has degenerated into an anti-philosophical praxis; it owes its strong impact to the ruthlessness with which, in Prague and elsewhere, it bans philosophy and suppresses its medium, freedom of thought. Because of this, official 'philosophy' in its current state has nothing to contribute regarding problems of international right. If we refuse to content ourselves with this, we will have to return to the point where great philosophy spoke its last words on the subject of 'international right', that is, to Kant. Recalling Kant's transcendental deduction of international right serves neither the historical interest of a detailed interpretation of a neglected part of his philosophy, nor the intention to take modern thought back to a position that is no longer suited to the problems of the twentieth century. It is meant as a mere preparatory work for the attempt to formulate the philosophical problem of a framework of international right in its current form; for we will certainly not find answers to the questions of today by regressing beyond the level of reflection that Kant already reached two hundred years ago.

5.1 The Systematic Locus of Kant's Transcendental Deduction of International Right

Kant's transcendental grounding of international right has been forgotten, and was never developed any further in philosophy because no one after him was able to reach the universality of his horizon. 'The horizon is the congruence between the limits of our cognition with the ends of mankind'[3]: that is how Kant defines the area to which human thought must shift in order to grasp the possibility and the aims of a framework of international right. The cognition of the *limits* of our cognition, that is, the *critique* of human reason, is assumed here as much as the cognition of its central goals: moral philosophy and transcendental teleology. Kant predicates his doctrine of international right on the 'world concept' of philosophy: 'philosophy' is 'the science of the relation of all cognition to the essential ends of human reason (*teleologia rationis humanae*).[4] Within this conception of philosophy, the transcendental deduction of public right (i.e. constitutional right, international right and cosmopolitan right) constitutes the climax of the philosophy of *history*. It is too little known that Kant's grounding of the philosophy of history in its principles and its structure went far enough for us to reconstruct a *Critique of Historical Reason* (if this formulation is permissible) with a high degree of precision. The essays dedicated to this subject belong to two phases of his work. The first phase is from 1784 to 1786; it is thus positioned between the publication of the first (1781) and second (1787) editions of the *Critique of Pure Reason*.[5] The second phase is from 1793 to 1798; this locates it after the publication of the *Critique of Practical Reason* (1788) and the *Critique of the Power of Judgement* (1789).[6] It is widely believed that Kant still changed, or at least modified, his philosophical position several times after the publication of the *Critique of Pure Reason*. Contrary to this opinion, closer analysis reveals that he

---

[3] Immanuel Kant, *Lectures on Logic*, ed. and trans. J. Michael Young (Cambridge University Press, 2004), p. 272.

[4] Immanuel Kant, *Critique of Pure Reason*, ed. and trans. Paul Guyer and Allen W. Wood (Cambridge University Press, 1998), p. 695.

[5] 1784: 'Idea for a Universal History with a Cosmopolitan Purpose'; 'An Answer to the Question: What Is Enlightenment?'
1785: Reviews of J. G. Herder's 'Ideas on the Philosophy of the History of Humanity', Part 1.2
1786: 'The Presumed Beginning of Human History'; 'What Does it Mean to Orient Oneself in Thinking?'

[6] 1793: 'On the common saying: That may be correct in theory, but it is of no use in practice'
1795: 'Perpetual Peace'
1798: 'The Contest of Faculties'.

worked out his philosophy of history with rigorous consistency and clear purpose; the outline established in the first essay was structured with greater detail, nuance and precision in its later elaboration, but it remained in place. The systematic outline is maintained, and already in 1784 it presupposes the transcendental teleology that was only fully developed in the *Critique of the Power of Judgement*, but whose foundations were already sketched in the appendix to the 'Transcendental Dialectic' and the 'Transcendental Doctrine of Method' from the *Critique of Pure Reason*. In all its stages of development, Kant's philosophy of history rests on the unity of the system of the 'legislation of human reason', which he derived from the 'original idea of a philosophy of pure reason' (B875) in the chapter 'The Architectonic of Pure Reason' (B860ff.) and never abandoned.[7]

Kant places the doctrine of international right in a systematic context that is not easily reconciled with the traditional schemata of philosophical, juridical or political thought, and therefore requires special explanation. Its systematic locus, as stated earlier, lies not in the philosophy of law, which can therefore explicate public right only incompletely, but in the philosophy of history. In the latter context, however, it occupies a unique position. For Kant's entire philosophy of history is based on the intention of uncovering the conditions of possibility for a future framework of international right. Kant's philosophy of history, then, focuses above all on the transcendental deduction of the legal situation between peoples that Kant terms 'eternal peace'. It follows from this that we must look for the transcendental principles of international right in the philosophy of history. The problem targeted by his philosophy of history is formulated in the first sentence of the essay 'Idea for a Universal History with a Cosmopolitan Purpose': 'Whatever conception of the freedom of the will one may form in terms of metaphysics, the will's manifestations in the world of phenomena, i.e. human actions, are determined in accordance with natural laws, as is every other natural event.'[8]History, then, is the realm in which freedom appears in nature.

Kant had shown in the *Critique of Pure Reason* that nature and freedom can be thought together without contradictions; but the problem of how they are intertwined remained unsolved. In the *Critique of Practical Reason* he had presented the *principle* and *inner* possibility of freedom 'with metaphysical intention'; but it was left open how freedom is meant to *appear* in nature. If history 'examines the free exercise of the human will *on a large scale*',[9] if it is to show how the appearance of freedom, as a natural occurrence, is determined by universal laws of nature, then the philosophy of history must be capable of showing the possibility of the *transition* from thinking

[7] There is insufficient space here for a comprehensive justification of this thesis; I have worked one out, but it is not yet ready for printing. The outline of problems offered here must, even for its specialized topic – the connection between the theory of international right and the philosophy of history – content itself with a small selection of supporting material, and refrains from presenting the interpretations on which it relies. It was also necessary to dispense with a discussion of divergent views.

[8] Immanuel Kant, 'Idea for a Universal History with a Cosmopolitan Purpose', in *Political Writings* ed. H. S. Reiss and trans. H. B. Nisbet (Cambridge University Press 1991), p. 41.

[9] Ibid.

in accordance with principles of nature to thinking in accordance with principles of freedom (*Critique of the Power of Judgment*, Introduction B, XX). It must mediate between the critique of pure (theoretical) reason and the critique of pure practical reason. But the 'universal laws of nature' that determine human *actions* are, in Kant's outline, identical to the laws of *development* in this particular species of organic beings, namely, the living beings that are gifted with the faculty of reason. The natural development of organic beings can neither be cognized nor understood according to the rules of understanding laid out in the *Critique of Pure Reason*. In order to cognize the laws of animate nature, the reflecting power of judgement must follow maxims that do not come from the rules of the understanding, but are derived from a transcendental synthesis of the principles of reason with the productive faculty of imagination. It must imagine nature *as if* it were determined by teleological principles. The cognition of the universal laws of nature followed by human action in history, then, presupposes the transcendental teleology on which Kant already draws in 'Idea for a Universal History'. It presupposes the necessary rational idea that there *must* be 'a ground of the unity of the supersensible that grounds nature with that which the concept of freedom contains practically'.[10] The process of the transition from freedom into nature, i.e. into existence in time, is history. The *presentation* of this transition is right, which, where it obeys a principle of reason, belongs to the sphere of freedom, but, where it regulates the behaviour of empirical individuals in society as positive right, belongs to the sphere of appearance, and thus nature. 'Presentation' here means that this transition directly manifests itself in every legal norm and legal act. The shared theme of the philosophy of history and the philosophy of public right is therefore not pure (practical) reason as such, but rather the *existence* of reason in time. The transcendental central question for Kant's philosophy of history and public right is therefore this: how is the existence of reason and freedom possible in nature? What are the conditions of possibility under which the intelligible essence of reason and freedom can *appear* in nature through a particular species of empirical living beings?

It follows from this that the transcendental deduction of public right (and thus of right as such) cannot be derived one-sidedly from either the principles of freedom or the principles of nature. International right requires a twofold justification. On the one hand, it must be shown how right emerges from the universal principle of right, that is, from a rational idea, in order to verify that the framework of international right is in agreement with moral principles, and that reason must therefore demand the development of a framework of international right for the realization of its freedom. On the other hand, it is necessary to cognize and transcendentally justify a law of nature with which one can explain why a particular species of beings spreads over the entire Earth and organizes itself in the forms of public right. Kant carries out the first part of the deduction, the derivation of public right from the universal principle of right, in the 'Metaphysical First Principles of the Doctrine of Right', the first part of the *Metaphysics of Morals*. The second part, the derivation of public right

10 Immanuel Kant, *Critique of the Power of Judgment*, trans. Paul Guyer and Eric Matthews (Cambridge University Press, 2000), p. 63.

from a natural process, is the task of the philosophy of history, which must in turn, for the sake of consistency, have two parts: the first is history as a natural process and the second is history as a process of freedom. That the appearance of reason and freedom in nature must take the form of a *process* is transcendentally justified in Kant's philosophy of history. The idea of a targeted process, however, is only possible within the framework of a teleological explanation of nature; hence there is no transition between nature and freedom without teleology. Kant refers to a targeted process as 'progress'. If progress is not possible, then neither is international right.

That right, being an outward appearance of reason, belongs both to the sphere of nature and the sphere of freedom follows from the definitions that Kant uses in the 'Metaphysical First Principles of the Doctrine of Right': 'All duties are either *duties of right* (*officia iuris*), that is, duties for which external lawgiving is possible, or *duties of virtue* (*officia virtutis sive ethica*), for which external lawgiving is not possible.'[11] The term 'external lawgiving' means that the law can forbid me from violating its legislation through my outward actions. It cannot, however, compel me to adopt the maxims of its legislation as maxims for my own actions based on my spontaneous will. Right is therefore connected to the licence to coercion, whereas no one can be coerced into virtue. This distinction between virtue and right goes back to the transcendental-philosophical distinction between the 'thing in itself' and 'appearance'. In the doctrine of duties, the human being shows itself from a dual perspective. Firstly, it shows itself 'in terms of its capacity for freedom, which is wholly supersensible, and so too merely in terms of its *humanity*, its personality independent of physical attributes (*homo noumenon*)';[12] secondly, it reveals itself as a subject burdened with physical attributes, that is, as a human (*homo phaenomenon*).[13] The term 'humanity' thus refers to the intelligible being and the term 'human being' to the empirical appearance of that same being. The duties of right are duties of the empirical human being in society; the duties of virtue are duties of the same human being, 'in terms of its capacity for freedom, which is wholly supersensible'. Kant's 'universal doctrine of right' can only be understood if one interprets it as a central part of his system of transcendental philosophy. All of its definitions are attempts to answer the transcendental question of how it might be possible for the supersensible faculty of reason to relate to the external. The relationship between the 'Metaphysical First Principles of the Doctrine of Right' and the doctrine of right itself is thus comparable in its line of questioning to that between the 'Metaphysical First Principles of Natural Science' and physics. Its topic is not the doctrine of right as such, but rather the question of how it is at all possible for a doctrine of right to live up to the theoretical demands laid out by Kant and later suppressed. It is sufficient to study the first paragraph of the first part on private right in order to assure oneself of the precise correspondence between the line of questioning developed here and the line of questioning in the *Critique of Pure Reason*.

---

[11] Immanuel Kant, *The Metaphysics of Morals, Revised Edition*, trans. Mary Gregor (Cambridge University Press, 2017), p. 35.

[12] Ibid.

[13] Ibid.

The side of right – positive right, if one interprets it in such a one-sided fashion – that relates to the appearing existence of empirical individuals manifests itself in this principle: 'Strict right can also be represented as the possibility of a fully reciprocal use of coercion that is consistent with everyone's freedom in accordance with universal laws.'[14] We refer to this principle in order to make clear the necessity of a twofold derivation of right. On its intelligible side, this principle follows from the 'universal principle of right': 'Any action is *right* if it can coexist with everyone's freedom in accordance with a universal law, or if on its maxim the freedom of choice of each can coexist with everyone's freedom in accordance with a universal law'.[15] But this principle, which will be elucidated later, does not yet explain the transcendental possibility of coercion. For it is a modification of the categorical imperative relating to the sphere of right. It thus originates from a principle that is the highest moral principle because it contains the transcendental explanation for the possibility of reason and freedom within itself. 'Strict right', on the other hand, is defined by Kant as 'that which is not mingled with anything ethical' and 'requires only external grounds for determining choice; for only then is it pure and not mixed with any precepts of virtue. Only a completely external right can therefore be called *strict* (right in the narrow sense).'[16] The definition of strict right thus contains the transcendental formula for the complete externalization (manifestation as appearance) of reason. The clarity with which Kant moves into the sphere of nature through this definition is evident from the note on §E of the introduction to the doctrine of right, which we must not pass over even in this abridged presentation, for it demonstrates the rigorous consistency of transcendental execution:

> The law of a reciprocal coercion necessarily in accord with the freedom of everyone under the principle of universal freedom is, as it were, the *construction* of that concept, that is, the presentation of it in pure intuition a priori, by analogy with presenting the possibility of bodies moving freely under the law of the *equality of action and reaction*. In pure mathematics we cannot derive the properties of its objects immediately from concepts but can discover them only by constructing concepts. Similarly, it is not so much the *concept* of right as rather a fully reciprocal and equal coercion brought under a universal law and consistent with it, that makes the presentation of that concept possible.[17]

So we see that, although the 'Metaphysical First Principles of the Doctrine of Right', in keeping with their title as part of the *Metaphysics of Morals*, only describe the one side of the transcendental deduction of right, namely its deduction from the principle of freedom, the indispensability of a transcendental justification of the possibility of *external* right, that is, the derivation of human action in nature from universal laws of nature, is immediately discernible in the attributes defined there. Our approach is therefore to begin by retracing the derivation of public right from

---

[14] Ibid., p. 28.

[15] Ibid., p. 27.

[16] Ibid., p. 28.

[17] Ibid., p. 29.

the natural process; then, in a third part, we will attempt to outline the process of freedom.

5.2 The Deduction of International Right from Natural Principles

As we saw, international right requires transcendental-philosophical justification because it deals with the question of how the *existence* of reason might be possible, that is, how reason and freedom can appear in the empirical world. It emerged from the 'Metaphysical First Principles of the Doctrine of Right' that the *concept* of right alone is not sufficient for this. Kant concludes from the *Critique of Pure Reason* that the possibility of right only becomes apparent if the *concept* of right is accompanied by a *construction* of this concept in the intuition. All that can be given to us in the intuition is 'nature' in the broadest sense of the term. In right, nature manifests itself as coercion: '[...] one can locate the concept of right directly in the possibility of connecting universal reciprocal coercion with the freedom of everyone.'[18] A twofold deduction of the possibility of right is therefore required: deduction from the principle of freedom and deduction from principles of nature. Because right in all its branches presupposes a society, however it might be structured, to perform this coercion into right, the task of deducing right from natural principles grows into the deduction of *public* right. For the sake of methodological clarity, this deduction must be carried out in such a way that no recourse to human *reason* is required. It must suffice that the attribute distinguishing the human being from other living beings is *understanding* [*Verstand*]. Certainly reason and nature are in constant interaction in the process of history. But the methodological task of deducing the possibility of right from natural laws is only fulfilled if positive public right can even be derived purely from natural principles, excluding the principles of reason. In keeping with this methodological intention, Kant declares: 'As hard as it may sound, the problem of setting up a state can be solved even by a nation of devils (so long as they possess understanding).'[19]

Right, as we saw, rests on 'the possibility of connecting universal reciprocal coercion with the freedom of everyone.' Therefore, right can only be derived from natural laws if such a connection between nature and freedom is possible *in* nature. In 'Idea for a Universal History', he seeks to show this by inserting, between his rigorously deterministic grounding of a theory of objective natural cognition according to the principles of reason (the 'transcendental analytic' in the *Critique of Pure Reason*) and his theory of a natural cognition according to the principles of transcendental teleology (the 'doctrine of method' in the *Critique of Pure Reason* and the *Critique of*

[18] Ibid., p. 28.

[19] Immanuel Kant, 'Perpetual Peace', in *Political Writings*, p. 112.

the Power of Judgment), a transcendental reflection that is meant to outline the possibility of such a transition. Though ostensibly only mentioned in passing, this reflection is of great significance for the systematics of right, for it shows how 'universal reciprocal coercion' and the 'freedom of everyone' can be directly connected. It is therefore necessary to quote these words:

> Thus marriages, births, and deaths do not seem to be subject to any rule by which their numbers could be calculated in advance, since the free human will has such a great influence upon them; and yet the annual statistics for them in large countries prove that they are just as subject to constant natural laws as are the changes in the weather, which in themselves are so inconsistent that their individual occurrence cannot be determined in advance, but which nevertheless do not fail as a whole to sustain the growth of plants, the flow of rivers, and other natural functions in a uniform and uninterrupted course.[20]

Hence Kant already recognized the importance of statistics for the theory of unorganized movement ('the changes in the weather, which are so inconsistent') a century before Boltzmann; in addition, he already discovered that statistics allows us to reconcile the constancy of natural laws with the indeterminacy of individual processes. He discovers in the laws of statistics a type of rule that can be applied both to human actions (marriages, births and deaths) and purely physical processes (unorganized movement). Since right can also be presented as a system of laws that endures even though the actions of the individual subjects are not strictly determined by these laws, there is a formal analogy between the laws of statistics and those of right that shows how laws of right can be understood as a particular modification of natural laws – which does not mean that this finding alone is sufficient to grasp the full nature of the law of right. All further convergences between natural law and the law of right result from the fact that, according to Kant, biology – that is, the theory of the development of organisms – cannot do without the transcendental hypothesis of teleological natural principles. So there must be natural laws governing a targeted development. Under this precondition, as Kant states with initial caution, one can hope that 'what strikes us in the actions of individuals as confused and fortuitous may be recognized, in the history of the entire species, as a steadily advancing but slow development of man's original capacities.'[21] In the transcendental grounding of the philosophy of history, Kant attempts to solve the problem thus posed; it is identical to the transcendental deduction of the appearing side of right. Kant already laid down the outline for his 'construction' of history as a *natural process* in 'Idea for a Universal History', in a form that was maintained in all later disquisitions. We find this outline in the first four 'propositions' and the accompanying explanations. Interpreting these sections also gives us a solid foundation for understanding his theory of international right. The 'First Proposition' reads: 'All the natural capacities of a creature are destined sooner or later to be developed completely and in conformity with their end.'[22] This, as Kant says in the explanation (three years before the publication of the second edition of the *Critique of Practical Reason*!),

[20] Kant, 'Idea for a Universal History with a Cosmopolitan Purpose', p. 41.

[21] Ibid.

[22] Ibid., p. 42.

is a foundation of the 'teleological doctrine of nature' and applies to all animals. The transcendental justification for this proposition is hinted at briefly: 'For if we abandon this basic principle of nature, we are faced not with a law-governed nature, but with an aimless, random process, and the dismal reign of chance replaces the guiding principle [*Leitfaden*] of reason.'[23] The phrase 'guiding principle of reason' refers to a unjustly neglected passage in the *Critique of Pure Reason* in which Kant explicitly states that the 'completion of the critical business of pure reason'[24] is not the transcendental deduction of pure concepts of the understanding, but rather the 'transcendental deduction of all the ideas of speculative reason'.[25] Kant shows that, in parallel with the 'schematicism of the pure concepts of the understanding', we also require a 'schema of the regulative principles for the systematic unity of all cognitions of nature.'[26] Without the regulative principles posited with this schema, there would be no basis for the idea of the 'systematic unity of the manifold of empirical cognition in general'[27] that underlies the 'transcendental analytic'. That is why, in the *Critique of Pure Reason* too, the entire cognition of nature only finds its grounding in transcendental teleology.

A schema is characterized by the fact that 'no object is given, not even hypothetically'. It 'serves only to represent other objects to us, in accordance with their systematic unity, by means of the relation to this idea, hence to represent these objects indirectly.'[28] It is in this sense that Kant sets up the principle on which the entire 'transcendental deduction of the ideas of speculative reason' is based. It reads as follows:

> Thus I say the concept of a highest intelligence is [...] a schema, ordered in accordance with the conditions of the greatest unity of reason, for the concept of a thing in general, which serves only to preserve the greatest systematic unity in the empirical use of our reason, in that one derives the object of experience, as it were, from the imagined object of this idea as its ground or cause. Then it is said, e.g., that the things in the world must be considered as if they had gotten their existence from a highest intelligence.[29]

This is the 'guiding principle of reason' that transcendentally legitimizes the 'First Proposition' in the text on the 'Idea for a Universal History'. As the deduction of right is the transition from the sphere of nature to the sphere of freedom, however, we should not restrict our investigation of this principle to the *Critique of Pure Reason*, but should also consult the *Critique of Practical Reason*, whose results, as can be shown, Kant already presupposes in the *Critique of Pure Reason*. There we find that the principle of transcendental teleology is far more than a mere schema of theoretical reason:

---

[23] Ibid.

[24] Kant, *Critique of Pure Reason*, p. 605.

[25] Ibid., p. 606.

[26] Ibid., p. 607.

[27] Ibid., p. 606.

[28] Ibid., p. 605.

[29] Ibid., p. 606.

> Now, the concept of freedom, in so far as its reality is proved by an apodictic law of practical reason, constitutes the *keystone* of the whole structure of a system of pure reason, even of speculative reason; and all other concepts (those of God and immortality), which as mere ideas remain without support in the latter, now attach themselves to this concept and with it and by means of it acquire stability and objective reality, that is, their *possibility* is *proved* by this: that freedom is real, for this idea reveals itself through the moral law.[30]

If the concept of a highest intelligence is proved in its 'objective reality' (!)[31] by the fact that freedom is real, then the principle of a teleological observation of nature gains, through the *Critique of Practical Reason*, a transcendental justification that Kant could not yet allow himself to employ in the *Critique of Pure Reason*, but which gives his transcendental deduction of the ideas of reason an entirely different consistency. So the teleology of the first historico-philosophical text is not a regression to pre-critical thinking, as was believed; rather, Kant can draw on the transcendental proof that without the 'guiding principle of reason' to which he refers here, reason cannot be reason and the understanding cannot be the understanding. It is hardly surprising, then, that this claim does not appear for the first time here but already in the *Critique of Pure Reason*, in both a more general and a more specific form. This is the general form: 'Everything that nature itself arranges is good for some aim.'[32] Specified to refer to humans, this becomes: 'Everything grounded in the nature of our powers must be purposive and consistent with their correct use'.[33]

It was necessary to explain the justification of the 'First Proposition' in some depth because the entire transcendental deduction of right is based on it. In right, nature and freedom are linked. The transcendental justification of the natural process from which the appearing side of right emerges can only agree later with the deduction of the metaphysics of right from principles of reason if, before the two paths of deduction diverge, the approach refers back to the 'ground of the unity of nature and freedom'.[34]

The 'Second Proposition' in 'Idea for a Universal History' explains why the human being, unlike the other living beings, has a history:

> In man (as the only rational creature on earth), these natural capacities which are directed towards the use of his reason are such that they could be fully developed only in the species, but not in the individual.[35]

This claim is supported with an anthropological thesis that still seems modern today:

---

[30] Immanuel Kant, *Critique of Practical Reason*, ed. and trans. Mary Gregor (Cambridge University Press, 2015), p. 3.

[31] The concept of 'objective reality' has a different meaning here compared to the 'Transcendental Analytic' from the *Critique of Pure Reason*; it is not a slip of the tongue, however, for it has a precise and verifiable sense that we cannot discuss here for reasons of space.

[32] Kant, *Critique of Pure Reason*, p. 645.

[33] Ibid., p. 590.

[34] Kant, *Critique of the Power of Judgment*, p. 63.

[35] Kant, 'Idea for a Universal History with a Cosmopolitan Purpose', p. 42.

> Reason, in a creature, is a faculty which enables that creature to extend far beyond the limits of natural instinct the rules and intentions it follows in using its various powers, and the range of its projects is unbounded. But reason does not itself work instinctively, for it requires trial, practice and instruction to enable it to progress gradually from one stage of insight to the next.[36]

In keeping with the methodological intention of the transcendental deduction of the natural process, reason is here developed not on its own terms as reason, but is described in the same way a biological species with this faculty would appear to a scientific observer. From this perspective, the two most important properties of a living being that possesses reason are that (1) reason is the faculty to emancipate oneself from the mechanisms of instinct, and (2) reason is the faculty to accumulate experience, knowledge and insight from generation to generation. Thus it forces humans to go constantly beyond themselves. With living beings that are confined by the limits of their natural instincts, the purposiveness of the species can only be completed in certain individuals, for they cannot transcend themselves; with them there is no inheritance of acquired characteristics. The human being, by contrast, can pass on acquired experiences, knowledge and morals, and is thus directed to a line of development that gives each generation the possibility to go beyond the previous generations. That is why humans have history; that is why the natural capacities with which they are equipped can 'be fully developed only in the species, but not in the individual'. The difference between reason and instinct has empirical evidence, which Kant adduces here. But it can also be transcendentally justified through reflection on the principles of the faculty of reason. This justification is found in the *Groundwork of the Metaphysics of Morals* and the *Critique of Practical Reason*. We can therefore say that the 'Second Proposition', together with its explanation, outlines the conditions of possibility for a species of living beings to have history in nature. The heart of this transcendental deduction of history is the statement that for reason, 'the range of its projects is unbounded'. We will see later how this statement links history viewed as a natural process with history viewed as a process of freedom.

The 'Third Proposition' reads thus:

> Nature has willed that man should produce entirely by his own initiative everything which goes beyond the mechanical ordering of his animal existence, and that he should not partake of any other happiness or perfection than that which he has procured for himself without instinct and by his own reason.[37]

How does one transcendentally justify this proposition? Kant says at the start of his explanation, 'For nature does nothing unnecessarily and is not extravagant in the means employed to reach its ends.'[38] This statement, under the name 'lex parsimoniae', is one of the basic principles of Kant's transcendental teleology. Kant already says in the introduction to the *Critique of the Power of Judgment* that it can be derived from the transcendental principle of the formal purposiveness of nature as an a

[36] Ibid.

[37] Ibid., p. 43.

[38] Ibid.

priori maxim.[39] If one combines the *lex parsimoniae* with the 'Second Proposition', the 'Third Proposition' follows as the conclusion. The philosophical scope of the step taken here becomes clear if one combines the principle that reason's 'range of [...] projects is unbounded' with the principle that 'man should produce entirely by his own initiative everything which goes beyond the mechanical ordering of his animal existence'. For the combination of these two principles yields a universal theory of *production*. The human being is forced by its biological constitution to produce culture, and this production is a process that pushes each generation beyond the preceding one, such that no generation ever reaches the goal. In his explanatory remarks, Kant himself says that this picture of human history is 'disconcerting': 'for in the actual course of human affairs, a whole host of hardships awaits him.'[40] Herein lies his theory of work, which he interprets as the continued production of each respectively higher level of self-realization by the human species. What is disconcerting about the incessant process of work is that it is maintained by humans 'without their intention'. The gradual working out of reason and freedom cannot itself follow from reason and freedom; the self-realization of the human species takes place behind the backs of humans, as it were. How this might be possible is not yet clear from the principles developed so far; the explanation only emerges from the 'Fourth Proposition'. That is why Kant, in his explanation of the 'Third Principle', calls this process not only 'disconcerting' but also 'puzzling'. Assuming that this puzzle will ultimately be solved, however, we can now observe that the transcendental deduction of history as a natural process was carried out with rigorous consistency: the 'First Proposition' grounds the teleological explanation of all processes of development in nature. The 'Second Proposition' derives, from the specific capacities of the human species, the necessity for its development to take place in the form of history. The 'Third Principle' determines the structure of this history as a process of production with the goal of human self-realization. It is almost superfluous to add that this theory of history, mediated through Hegel, flowed into Marxism and thus took on world-historical significance.

Kant takes the most important step for the deduction of public right in the 'Fourth Proposition', which is meant to solve the puzzle posed in the 'Third Proposition':

> The means which nature deploys to bring about the development of innate capacities is that of antagonism within society, in so far as this antagonism becomes in the long run the cause of a law-governed social order.'[41]

To understand the meaning of this statement, one must be aware of the problem that Kant is seeking to solve. In his reflections on anthropology, Kant noted down, 'Perhaps all humans are disturbed in a certain sense (lunatic asylum of the universe)'.[42] In the essay 'Conjectural Beginning of Human History', he says: 'The

---

[39] Kant, *Critique of the Power of Judgment*, p. 69; regarding the transcendental deduction of this maxim, see *Critique of Pure Reason*, p. 598.

[40] Kant, 'Idea for a Universal History with a Cosmopolitan Purpose', pp. 43f.

[41] Ibid., p. 44.

[42] Immanuel Kant, *Akademie-Ausgabe*, vol. XV: *Reflexionen zur Anthropologie* (Berlin and Leipzig: de Gruyter, 1923), p. 211.

history of *nature* thus begins from good, for that is the *work of God*; the history of *freedom* from evil, for it is the *work of the human being.*'[43] Here Kant formulates an approach to a problem that would remain in his focus until the text 'Perpetual Peace'. The development of the human species for its own good, which is demanded by the principles of teleology, can only be shown by revealing a mechanism whereby nature uses the drive towards evil in order to 'work out' good. The term 'mechanism' should be understood in the strict sense of the word: Kant set himself the task of showing that without any intention of their own, even against their will, humans are forced by an irresistible force of nature to develop social and political orders that, once established, render reason and freedom possible.

It is in this context that the 'Fourth Proposition' should be interpreted. The antagonism of which Kant speaks lies in the fact that the human being has (1) 'an inclination to *live in society*' and (2) 'a great tendency to *live as an individual*'. Kant may have taken this classification of basic human drives, which points back to the Stoics, from any number of sources. One might think, for example, of Shaftesbury's theory that human inclinations are either social or self-interested. What distinguishes Kant from his philosophical predecessors, however, is that, contrary to the mentality of his century, he does not arrive at his ethics by establishing a harmony between the drive to preserve the species and the drive to self-preservation, but rather by placing reason, because its principle is freedom, in opposition to the natural 'inclination' of human drives. So there is a twofold antagonism: the antagonism between sociality and self-interest and the antagonism between both of these drives and reason, which is, however – *contrary to* the intention of these drives – meant to result from a mechanism set in motion by the antagonism itself.

We are not explaining the play of this antagonism from the text of the essay 'Idea for a Universal History with a Cosmopolitan Purpose', as we did with the preceding propositions, but with reference to 'Perpetual Peace', because this allows us to show at once how Kant derives public right from a natural mechanism in the context of his historico-philosophical conception. Kant proceeds from an analysis of the situation 'in which nature has placed the actors in her great spectacle'.[44]

> Nature's provisional arrangement is as follows. Firstly, she has taken care that human beings are able to live in all the areas where they are settled. Secondly, she has driven them in all directions by means of war, so that they inhabit even the most inhospitable regions. And thirdly, she has compelled them by the same means to enter into more or less legal relationships.[45]

This is meant to explain the origin of the initial conditions Kant presupposes for the derivation of international right. He had already named the first condition earlier in the text: that humans on Earth inhabit a globe, so 'they cannot disperse over an infinite area, but must necessarily tolerate one another's company.'[46] A second

---

[43] Immanuel Kant, 'Conjectural Beginning of Human History', in *Anthropology, History and Education*, ed. Robert B. Louden and Günter Zöller (Cambridge University Press, 2007), p. 169.

[44] Kant, 'Perpetual Peace', p. 109.

[45] Ibid., p. 110.

[46] Ibid., p. 106.

condition lies in the fact that the whole Earth is populated. This in turn presupposes that humans can live even in inhospitable parts of the Earth, and that they were forced by necessity to settle in those areas. This force, which can be derived mechanically from the antagonistic interplay of the drives to sociality and isolation, is war. But war also coerces them 'to enter into more or less legal relationships':

> Even if people were not compelled by internal dissent to submit to the coercion of public laws, war would produce the same effect from outside. For in accordance with the natural arrangement described above, each people would find itself confronted by another neighbouring people pressing in upon it, thus forcing it to form itself internally into a *state* so that it is equipped to encounter the other as a *power*.[47]

One sees how consistently Kant is at pains in his methodology to exclude any impact of morals and practical reason in his derivation of the state from the natural process. This does not mean it is negated; it means that the deduction of the state from principles of nature needs to be augmented by its deduction from principles of freedom.

The deduction of international right, by the nature of the matter, is far more complicated. 'The idea of international right presupposes the *separate existence* of many independent adjoining states'; it intends for them to form a 'federal union'.[48] It thus demands an alliance that 'involve[s] no sovereign authority (as in a civil constitution), but only an *association* (federation)'.[49] Because of 'the depravity of human nature', which 'is displayed without disguise in the unrestricted relations which obtain between the various nations',[50] it is difficult to solve the problem of showing a natural mechanism that uses this same depravity as a vehicle to bring about the reasonable state of 'perpetual peace' secured by international right. To understand Kant's deduction, one must incorporate the derivation of cosmopolitan right. The self-interest that keeps the 'inclination to *live in society*' in motion manifests itself in the international context as the '*spirit of commerce*', which 'sooner or later takes hold of every people, and it cannot exist side by side with war'.[51] It takes political form in financial power, which, 'of all the powers (or means) at the disposal of the power of the state, [...] can probably be relied on most'.[52] The spirit of commerce represents the impulse, while financial power and state power represent the masses to be moved. Politics can thus be described as a form of physics.

> Thus states find themselves compelled to promote the noble cause of peace, though not exactly from motives of morality. And wherever in the world there is a threat of war breaking out, they will try to prevent it by mediation, just as if they had entered into a permanent league for this purpose [...].[53]

[47] Ibid., p. 112 (translation modified).

[48] Ibid., p. 113.

[49] Kant, *The Metaphysics of Morals*, p. 125.

[50] Kant, 'Perpetual Peace', p. 103.

[51] Ibid., p. 114.

[52] Ibid.

[53] Ibid.

The 'inclination to *live in society*' that manifests itself as the spirit of commerce, however, is counteracted by the 'tendency to *live as an individual*'. States have the tendency to separate from one another. This tendency is not only explained by the origins of the state in war; it is explained by the defence against a danger that threatens all states. It is 'the desire of every state (or its ruler) to achieve lasting peace by thus dominating the whole world, if at all possible'; the federal system of international right is meant to prevent this, and nature 'uses two means to separate the nations and prevent them from intermingling – *linguistic* and *religious* differences'.[54] As the whole of public right, according to its three conditions as state right, international right and cosmopolitan right, is thus derived not from reason, but from human drives and the mechanisms set in motion by them, Kant can close by saying: 'In this way, nature guarantees perpetual peace by the actual mechanism of human inclinations.'[55] Because perpetual peace is simultaneously a postulate of reason, however, this would show that the goal of the natural process is identical to the goal that reason, according to its own principles, must set for itself.

The mechanism of nature outlined only briefly here is meant to demonstrate how nature guarantees that 'what man *ought* to do by the laws of his freedom (but does not do) will in fact be done through nature's compulsion, without prejudice to the free agency of man [...] in all three areas of public right'.[56] The essay 'Idea for a Universal History' also follows the first four 'propositions' with the derivation of 'a civil society which can administer justice universally'.[57] So the philosophy of history indeed contains the transcendental deduction of the coercive character of positive right from its function within a natural process. It is conspicuous that the natural mechanism described by Kant anticipates the figure of a law of motion that Hegel later developed as the 'cunning of reason'; one could therefore term it the 'cunning of nature'. But in Kant too, this cunning of nature is a cunning of the divine reason known in exoteric language as 'providence'. As far as I can see, the explanation of the 'Fourth Proposition' in 'Idea for a Universal History' is the first philosophical text to carry out a movement of thought that is 'dialectical' in Hegel's sense. The entire theory of the natural process has a dialectical structure. If, however, one observes more closely how dialectic enters the knowledge of nature, it transpires that it results not from empirically determinable facts, but from their transcendental justification. It is hardly necessary to emphasize what this means for our understanding of the origin and definition of the legitimate realm of dialectical thought.

---

54 Ibid., pp. 113f.

55 Ibid., p. 114.

56 Ibid., p. 112.

57 Kant, 'Idea for a Universal History with a Cosmopolitan Purpose', p. 45.

5.3 The Deduction of International Right from the Principle of Freedom

Kant's transcendental justification of international right is meant to show that the *existence* of reason is thinkable and possible *in* nature. The first part of this deduction – the derivation of the possibility of international right from the natural developmental history of the human species on this planet – uncovered a mechanism that allows us to see the overall direction of this development on the basis of natural principles. The development of this biological species must process as *history*: it must be a targeted process that, at unpredictable temporal intervals, can be presented as progress in the direction of a universal framework of international right. Because Kant, unlike the 'fantasists of reason' (Abbé St Pierre, Plato, Rousseau[58]), views political history with the sober eye of a pragmatist and constantly points out the unreason and depravity of humans, only a natural mechanism can 'guarantee' that the progress in history demanded by reason is actually possible. This is the meaning of the section heading 'On the Guarantee of a Perpetual Peace'.[59] It is the *conditio sine qua non* of progress that humans *must* move in the direction of a universal framework of international right, even without such an intention of their own. But the realization of reason and freedom can only arise positively from the principles of reason itself. Therefore, the transcendental deduction of international right from natural principles must be augmented by its deduction from principles of freedom. This second part of the transcendental deduction of international right does not, as one is perhaps inclined to think today, have the character of an 'ideological superstructure'; rather, it takes precedence in the architecture of the system of transcendental philosophy because it is only through it that we can understand how, in the course of history, a period of reason's self-knowledge could ensue in which it became possible to cognize the first part of the deduction.

In our presentation of the deduction of international right from the principle of freedom, we are proceeding in the same way as in the presentation of the natural process: our concern is not to lay out Kant's doctrine in its individual steps. Rather, we will concentrate on the task of interpreting the general outline of his doctrine with reference to its transcendental justification, that is, its connection to the whole system of transcendental philosophy. Only thus can we determine what the consequences might be if we were now forced, from a new historical perspective, to view Kant's entire philosophy as a past era in the history of reason.

(a) Our point of departure can then only be the principle Kant himself called the 'transcendental formula of public right': 'All actions affecting the rights of other human beings are wrong if their maxim is not compatible with their being

[58] Kant, *Reflexionen zur Anthropologie*, p. 210.

[59] Kant, 'Perpetual Peace', p. 108.

made public.'[60] The transcendental character of this formula comes from the method by which Kant derives it. He says:

> If, in considering public right as the jurists usually conceive of it, I abstract from all its material aspects (as determined by the various empirically given relationships of men within a state, or of states with one another), I am left with the *formal attribute of publicness*. For every claim upon right potentially possesses this attribute, and without it, there can be no justice (which can only be conceived of as *publicly knowable*) and therefore no right, since right can only come from justice.[61]

The cognitions that remain if one leaves aside all the matter of cognition, that is, all empirically given facts, and only considers the conditions of possibility for this givenness, are transcendental; they can be recognized a priori by reason. Kant therefore says of the 'transcendental formula of public right' that, 'like any axiom, it is valid without demonstration'.[62] This does not mean, however, that the a priori nature of this cognition, like the a priori nature of all other a priori cognitions, does not require a transcendental deduction. The second part of the deduction of public right, its derivation from the principle of freedom, succeeds when we see through the deduction of the 'transcendental formula of public right', without which 'there can be no justice [...] and therefore no right'.

The systematic context we enter with this question becomes transparent if we note that the formula posited by Kant in the text 'Perpetual Peace' recapitulates an important section from the *Critique of Pure Reason* in concentrated form. For the principle of the publicness of right is only a special case of the general principle of publicness of reason that Kant formulates in the *Critique of Pure Reason* at the start of the section 'The discipline of pure reason with regard to its polemical use':

> Reason must subject itself to critique in all its undertakings, and cannot restrict the freedom of critique through any prohibition without damaging itself and drawing upon itself a disadvantageous suspicion. Now there is nothing so important because of its utility, nothing so holy, that it may be exempted from this searching review and inspection, which knows no respect for persons. The very existence of reason depends upon this freedom, which has no dictatorial authority, but whose claim is never anything more than the agreement of free citizens, each of whom must be able to express his reservations, indeed even his veto, without holding back.[63]

On first reading these lines, one is inclined to view them merely as an expression of a basic political conviction; nonetheless, one cannot fail to see that the demand for freedom of thought has a clearly defined place in Kant's philosophy. It is indicated by the words, 'The very existence of reason depends upon this freedom.' The existence of reason is the central problem of the philosophy of history; its *conditio sine qua non* is public right. The general demand for

---

[60] Ibid., p. 126.

[61] Ibid., p. 125.

[62] Ibid., p. 126.

[63] Kant, *Critique of Pure Reason*, p. 643.

publicness of thoughts is therefore in keeping with the special demand for publicness of right as outlined in the 'transcendental formula of public right'.

So the section begins with these words: 'Reason must subject itself to critique in all its undertakings.' This statement contains a direct reference to the famous passage from the preface to the first edition of the *Critique of Pure Reason* in which Kant explains the title of the work:

> Our age is the genuine age of critique, to which everything must submit. *Religion* through its *holiness* and *legislation* through its *majesty* commonly seek to exempt themselves from it. But in this way they excite a just suspicion against themselves, and cannot lay claim to that unfeigned respect that reason grants only to that which has been able to withstand its free and public examination.[64]

One can show that this passage contains the conceptual outline for the text 'What is Enlightenment?'. But 'man's emergence from his self-incurred immaturity'[65] is described there as a *historical* process; alongside the essay 'Idea for a Universal History', written the same year (1784), this essay stands at the beginning of Kant's work on the philosophy of history. So the chronology also confirms that, in parallel with his derivation of history from the natural process ('Idea for a Universal History'), Kant examined the problem of deriving a *reasonable* history from the principle of freedom ('What is Enlightenment?'). The two derivations converge in the concept of the 'existence of reason'. When Kant says that the 'claim' of reason is 'never anything more than the agreement of free citizens', he is referring to the political precondition for the existence of reason: the republican constitution. According to Kant, however, this is the only form of constitution that is in full agreement with the 'transcendental formula of public right'. Thus ostensibly heterogeneous pieces of Kant's philosophy are seamlessly joined.

The systematic relevance of this initially merely superficial observation only comes to light, however, when one considers that the freedom which is supposed to be secured by public right, and on which the existence of reason rests, is, both in section B 766f. of the *Critique of Pure Reason* and in the note on section A XI, brought into direct connection with the central concept of transcendental philosophy as such, namely the concept of a 'critique' of reason. The philosophical background of the transcendental formula of public right reveals itself if we ask how the *critique* of reason is connected to the *existence* of reason. It will become apparent that the second part of Kant's philosophy of history is nothing other than a development of this connection. To understand this, we must address a frequently neglected side of Kant's concept of 'critique'.

In the preface to the second edition of the *Critique of Pure Reason*, Kant states that he wants to set in motion a 'revolution in the way of thinking' with

---

[64] Ibid., pp. 100f.

[65] Immanuel Kant, 'An Answer to the Question: What Is Enlightenment?', in *Political Writings*, p. 54.

this work.[66] He sees this revolution as a 'Copernican turn' that is unique in the historical process, and in which reason attains self-knowledge. This act of emancipation opens up the secure path of a progress in rational cognition that can be slowed, disturbed or interrupted, but no longer revoked, and takes hold of all spheres of human existence. The earlier phases of history now seem mere prehistory; with the Copernican turn, history begins *in* history, namely the history of autonomous reason. In so far as the *Critique of Pure Reason* brings about a revolution in history, the concept of 'critique' itself has a historico-philosophical meaning; in this sense it corresponds to Kant's philosophical concept of 'enlightenment'. How can we justify the fact that Kant identifies his 'metaphysics of metaphysics' (letter to Marcus Herz written after 11 May 1781) with the political concept of revolution?

The title *Critique of Pure Reason*, in addition to the overall meaning that encompasses the entire work, has a particular connection to the second division of 'The Transcendental Logic', namely 'The Transcendental Dialectic'. In 'The Transcendental Dialectic', Kant shows that all cognition of God, the world and the soul drawn on by traditional metaphysics (and thus also theology) rests on a 'transcendental illusion [*Schein*]'. In negative terms, critique in the form of Kant's dialectic has the task of penetrating and dissolving this illusion; it thus carries out a destruction of classical metaphysics. Kant initiated the period in the history of philosophy in which thought repeatedly found itself forced to begin its work with an act of destruction; there is a direct path from Kant's transcendental dialectic to Marx's destruction of idealism, the destruction of all previous metaphysics, religion and morality in Nietzsche's doctrine of the 'genealogy of morals', and the destruction of the history of ontology in Heidegger's *Being and Time* or Adorno's *Negative Dialectics*. Kant was aware that dissolving the transcendental illusion would inevitably also send shockwaves through the political and ecclesiastical hierarchies that underpinned the edifice of the European order before the French Revolution; for this calls into question the metaphysics under whose protection '*religion* through its *holiness* and *legislation* through its *majesty*' seek to evade the critique of reason.

But the destructive and thus seemingly negative side is only the exterior surface of critique. For the dissolution of the transcendental illusion directly uncovers the truth, since only cognition of the truth provides the criteria that allow us to expose the illusion as an illusion. The positive side of critique lies in showing that the *truth* of possible rational cognition presupposes its *freedom*, that this freedom of reason is reconcilable with the principles of the understanding of causality that are restricted to the sphere of appearance, and that practical reason therefore takes absolute precedence over theoretical reason. 'Thus I had to undo *knowledge* in order to make room for *faith*; and the dogmatism of metaphysics, i.e., the preconception that one can make progress in metaphysics without a critique of pure reason, is the true source of all

[66] Kant, *Critique of Pure Reason*, p. 108.

unbelief conflicting with morality, which unbelief is always very dogmatic.'[67] The destruction of traditional metaphysics is only the precondition for the 'Copernican turn', which gains autonomy through the discovery of the primacy of practical reason in an act of critical self-knowledge. This turn deserves the political name 'revolution' because the transcendental dialectic pulls the carpet from under dogmatism, on which all historical forms of despotism are based. Reason can only fend off despotism if it 'sees reason', that is, if it becomes philosophically reflected reason. For Kant, this is the only true revolution. He rejected political revolution, on the other hand. He says:

> A revolution may well put an end to autocratic despotism and to rapacious or power-seeking oppression, but it will never produce a true reform in ways of thinking. Instead, new prejudices, like the ones they replaced, will serve as a leash to control the great unthinking mass.[68]

After this general reminder of the context one must have in mind in order to understand the transcendental formula of public right, there are two questions left for us to resolve: (1) how does the revolution in the way of thinking relate to the transcendental justification of right according to principles of freedom? (2) How does it fit into the context of Kant's philosophy of history?

(b) Answering the first question requires a closer consideration of how to justify the formula transcendentally. Kant gives the following indication: 'This principle should be regarded not only as *ethical* (i.e. pertaining to the theory of virtue) but also as *juridical* (affecting the right of man).'[69] So the principle, in keeping with the deduction of public right from the autonomy of reason, is meant to unify the doctrine of morals and the doctrine of right. The doctrine of morals is the doctrine of the maxims and goals that reason is meant to set itself out of freedom; its system stems purely from reason. The concept of right is likewise a pure concept, but in 'application to cases that come up in experience'.[70] Hence the doctrine of morals is concerned with freedom as such, and the doctrine of right with freedom in the realm of experience. The doctrine of morals is concerned with pure reason, while the doctrine of right deals with the existence of reason.

If the transcendental formula of public right is to be viewed as 'juridical', it presupposes the universal principle of right as such: 'Any action is *right* if it can coexist with everyone's freedom in accordance with a universal law, or if on its maxim the freedom of choice of each can coexist with everyone's freedom in accordance with a universal law.'[71] The transcendental justification of this principle emerges if one compares it with the principle of freedom, the categorical imperative: 'Act only according to that maxim through which you

[67] Ibid., p. 117 (translation modified).

[68] Kant, 'An Answer to the Question: What Is Enlightenment?', in *Political Writings*, p. 55.

[69] Kant, 'On Perpetual Peace', p. 126.

[70] Kant, *The Metaphysics of Morals*, p. 3.

[71] Ibid., p. 27.

can at the same time will that it become a *universal law*.'[72] We will proceed by first explaining the transcendental meaning of the categorical imperative and arriving from there at a transcendental understanding of the universal principle of right.

The understanding of the categorical imperative has been obscured by its misinterpretation as an ethical principle and its framing within the lopsided opposition between 'formalism in ethics and the material value ethics' (Max Scheler). In Kant, the discipline we call 'ethics' is justified through the '*metaphysics* of morals'; this in turn requires justification through a 'metaphysics of metaphysics', namely the 'groundwork of the metaphysics of morals'. The place of the categorical imperative is in this groundwork. It is meant to transcendentally justify the general conditions of possibility for moral action as such, namely the possibility of the will. What Kant means by will is not a blind, irrational drive; 'the will is nothing other than practical reason'.[73] That is why the universal definition of the will, which must also always be kept in mind in the interpretation of the doctrine of right and the philosophy of history, is this: 'Every thing in nature works according to laws. Only a rational being has the capacity to act *according to the representation* of laws, i.e. according to principles, or a *will*.'[74] The objects that *appear* to us in nature are governed mechanically by laws according to the principle of causality. We say of a rational being that it can govern itself, and thus has a will, because the laws governing reason do not determine reason mechanically. The faculty of reason must produce of its own accord the law according to which reason is able to grasp itself as reason, and must make this law its task in such a way that it can act according to it. This faculty of reason constitutes its freedom, to which it empowers itself through the representation of its law. 'The representation of an objective principle in so far as it is necessitating for a will is called a command (of reason), and the formula of the command is called *imperative*.'[75] The 'representation of an objective principle' is the act of reason whereby it renders possible its own character to be will. So the word 'command' refers, transcendentally speaking, not to a directive from without or above obliging its recipient to act morally; rather, it refers to the ontological constitution of a living being to which its own being is not given, but rather assigned as a task.[76] This is the general ontological constitution of the will. It is, as Kant shows in his doctrine of transcendental ideas the ontological constitution of reason as such. 'The formula of the command is called *imperative*'; the term 'imperative'

---

[72] Kant, *Groundwork of the Metaphysics of Morals*, p. 34.

[73] Ibid, p. 27.

[74] Ibid., p. 26.

[75] Ibid., p. 27.

[76] This statement can be explained and justified in Kantian terms by noting that already in the *Critique of Pure Reason*, the 'nature' of the subject is outlined through a web of trans-categorial determinations of being that do not have the form of objective judgements, but follow regulative principles.

thus refers to the abstract structure of this ontological constitution. One could therefore encapsulate the meaning of the categorical imperative in a simple formula: if you will, you must act in such a way that you can will. Will as such, whatever form it takes, has the universal structure that it must first empower itself to itself. Only by empowering itself to itself is it free. If one asks *how* it empowers itself to itself, the answer is this: through 'the representation of an objective principle', that is, through reason.

Therefore, the categorical imperative does not have the meaning of a moral demand; it is a transcendental principle. 'It concerns not the matter of the action or what is to result from it, but the form and principle from which it does itself follow.'[77] The action follows from the will. The 'form' of the will is the form of reason, the principle of the will is freedom. Reason and freedom are the conditions of possibility of the will. If the categorical imperative directs itself at reason and freedom, it directs itself at the conditions of possibility of action that emerges from a free will. But the concepts of reason and freedom refer to the ontological constitution of the subject of a free action; they are, if the phrase is suitable, the transcendental determinations of being of the subjectivity of the subject. The will is the *essence* of the subject of action. Hence the categorical imperative has an 'ontological' meaning in Kant. It is the formula for the being of a subject that *is* by *willing* the freedom of its own will.

From this emerges the meaning of formulation of the categorical imperative: 'Act only according to that maxim through which you can at the same time will that it become a universal law.' The command to 'act only according to that maxim' is directed at the empirical subject, for only an empirical subject can act. This follows from the definition of a maxim: 'A *maxim* is the subjective principle for acting, and must be distinguished from the *objective principle*, namely the practical law.'[78] The categorical imperative thus means: act only according to that subjective principle through which you can at the same time will that it become an objective principle for every rational being. This formula corresponds precisely to what Kant, in the note to A 117 in the *Critique of Pure Reason*, calls 'the absolutely first and synthetic principle of our thinking in general'.[79] That is, translated into direct speech: all distinct empirical consciousness must be connected within a united self-consciousness. Just as thinking only cognizes the truth if it is capable of thinking what is true for *all* thinking beings, action is only free if it determines itself through the representation of a law that is binding for *all* rational beings, because it grounds in all of them the faculty to act out of freedom. The categorical imperative, then, commands nothing other than the transition from empirical to universal consciousness. It is the basic formula for the self-knowledge of reason as the faculty with which the rational living being frees itself for itself and its tasks.

[77] Kant, *Groundwork of the Metaphysics of Morals*, p. 30.

[78] Ibid., p. 34n2.

[79] Kant, *Critique of Pure Reason*, p. 237.

The universal principle of right has a very different meaning: 'Any action is *right* if it can coexist with everyone's freedom in accordance with a universal law, or if on its maxim the freedom of choice of each can coexist with everyone's freedom in accordance with a universal law.' Kant indicated the difference precisely:

> The doctrine of right dealt only with the *formal* condition of outer freedom [...], that is, with *right*. But ethics [...] provides [...] an end of pure reason which it represents as an end that is also objectively necessary'.[80]

Therefore, the law invoked in the universal principle of right should not be confused with the 'law' mentioned in the formula of the categorical imperative. The law of the categorical imperative is, as its second, explanatory version teaches, a 'universal law of nature'.[81] It refers to the universal condition of possibility for the nature of any rational being. By contrast, the 'universal law' that grounds the possibility of right is the constitution of a community structured in such a way that 'the freedom of choice of each can coexist with everyone's freedom'. Because Kant only stipulates the norm for *outer* freedom, the universal principle of right only applies to each person's freedom of *choice*, but not to the freedom of will as such, as the categorical imperative does. Yet this outer freedom must be guaranteed if a form of action resulting from true freedom, that is, from reason, is to be possible in society. For this reason, the universal principle of right formulates the condition of possibility for the *existence* of reason.

We have interpreted the 'universal law' as the constitution of a community, which means that we have understood the universal principle of right as the basic formula of *public* right. The justification for this follows from the passage in which Kant formulates this principle for the first time, and which also contains its transcendental justification. It is in the *Critique of Pure Reason*, in the chapter 'On the Ideas in General':

> A constitution providing for the *greatest human freedom* according to laws that permit *the freedom of each to exist together with that of others* [...] is at least a necessary idea, which one must make the ground not merely of the primary plan of a state's constitution but of all the laws too [...].[82]

This secures the constitutional meaning of the universal principle of right. Now the chapter 'On the Ideas in General' contains the grounding for the 'system of transcendental ideas'. It delineates the field for Kant's doctrine of the nature of reason. Thus the universal formula for the *existence* of reason is built into the doctrine of *pure* reason.

---

[80] Kant, *The Metaphysics of Morals*, p. 156.

[81] Kant, *Groundwork of the Metaphysics of Morals*, p. 34.

[82] Kant, *Critique of Pure Reason*, p. 397. It is worth noting that with this statement, Kant means to clarify the true intention of the 'Platonic republic', and that, in this context, he aptly refutes the misinterpretations of Plato's *Republic* that persist to this day.

We now return to the transcendental formula of public right: 'All actions affecting the rights of other human beings are wrong if their maxim is not compatible with their being made public.' We now see that this formula agrees with the universal principle of right; it expresses negatively what the latter expresses positively. But it also formulates what condition must be met so that 'the freedom of choice of each can coexist with everyone's freedom in accordance with a universal law'. This condition is publicness, for maxims that are not compatible with publicness violate freedom; if they did not, there would be no reason to keep them secret. The universal requirement of publicness is the justification for classifying state right, international right and cosmopolitan right as *public* right. This publicness is not only empirically grounded, however, but is defined in a 'transcendental formula', because it projects the universality of the law formulated by the categorical imperative into the outer sphere of society and the world of the state. Because the principles of reason are binding for all rational living beings as such, the appearance of reason in the empirical world necessarily presupposes the form of publicness. Reason attains existence in the medium of publicness; hence publicness of both right and thoughts is the medium in which reason and freedom realize themselves.

The question of the transcendental justification of public right must, Kant assumes, be rejected by jurists because their function is to apply existing laws, and the existing laws can only represent the principle of right in an incomplete form. The jurists have chosen not only the scales of the law but also the sword of justice as their symbols, and are always tempted to 'throw the sword into one of the scales if it refuses to sink'.[83] The framework of international right that is meant to secure perpetual peace is untenable unless it is grounded in reason and enables freedom. But this same fundamental condition contradicts the empirical constitutions of states, which emerge in the natural process from the self-interest of their rulers and peoples, and reflect these origins in the coercive character of constitutions, which conflict with the principle of right. From this dilemma, Kant concluded acutely yet also with bitter irony that the foundation of perpetual peace would be hidden in a 'hidden article on perpetual peace'. The article reads thus: 'The maxims of the philosophers on the conditions of possibility of public peace shall be consulted by states which are armed for war.'[84] 'The conditions of possibility of public peace' – this formulation refers in Kant's language to the transcendental grounding of public right; he reminds us that the reasonableness of a framework of right can only be justified by the principles of reason. Thus the problem of public right leads to the problem of the relationship between philosophy and politics.

> It is not to be expected that kings will philosophize or that philosophers will become kings; nor is it to be desired, however, since the possession of power inevitably corrupts the free judgement of reason. Kings or sovereign peoples (i.e. those governing themselves by egalitarian laws) should not, however, force the class of philosophers

---

[83] Kant, *Perpetual Peace*, p. 115.

[84] Ibid. (translation modified).

> to disappear or to remain silent, but should allow them to speak publicly. This is essential to both in order that light may be thrown on their affairs.[85]

(c) The second question to solve was this: how does the revolution in the way of thinking whereby reason attains self-knowledge and thus freedom, fit into the context of Kant's philosophy of history?

Unlike Herder, Kant demands of a 'philosophy of history of mankind' that it show 'logical precision in the definition of concepts' and 'careful distinctions and consistency in the use of principles'.[86] The first demand concerns the form of presentation, the second its systematic grounding. We have followed step by step how meticulously Kant worked out the central principles of his philosophy of history; now our analysis must confront the text in which Kant explicitly asks the transcendental question, 'But how is it possible to have history *a priori*?' This is found in the essay 'The Contest of Faculties' from 1798. The structure of this text confirms our findings about the systematic locus of Kant's philosophy of history: the transcendental question of the conditions of possibility of history has its place in the second section of the 'contest between the faculty of philosophy and the faculty of law'. Only here can he ask, 'Is the human race continually improving?',[87] for all progress in history, according to Kant, is tied to the evolution of public right into a universal framework of international right.

When Kant asks what are the conditions of possibility of history, he is interested not in past history but future history. History cannot be viewed philosophically in any other way as soon as one has understood that the form in which we interpret the past always depends on the form in which we anticipate the future. Kant therefore says at the outset:

> What sort of knowledge are we looking for? What we are seeking to know is a portion of human history. It is not a history of the past, however, but a history of future times [...].[88]

He excludes '*predictive*' history, which applies known laws of nature to determine such future events as solar or lunar eclipses in advance; he excludes a 'prognosticative' view of the coming time based on supernatural communication. Ultimately he excludes all particular events in the history of individual states and individual peoples, for history, according to Kant, deals with 'the *whole* of humanity (*universorum*), united in earthly society and distributed in national groups.'[89] Only in this strict universality does Kant allow the concept of history. Only within the horizon of this universality can he ask, 'Is the human

---

[85] Ibid.

[86] Immanuel Kant, 'Reviews of Herder's Ideas on the Philosophy of the History of Mankind', in *Political Writings*, p. 201.

[87] Immanuel Kant, 'The Contest of Faculties', in *Political Writings*, p. 177.

[88] Ibid.

[89] Ibid.

race continually improving?' Kant refers to the science that can answer this question as '*prophetic*' history.

In formulating his question in this way, he has already made a fundamental decision that becomes apparent in the heading of the fourth chapter: 'The problem of progress cannot be solved directly from experience'.[90] In analogy to the argumentation with which Hume showed that the universality of the law of causality cannot be empirically proved, Kant assumes that one cannot conclude from any historical progress, whatever its length, that it must continue indefinitely. Moreover, such a hypothesis would contradict our direct experience of history, which appears to teach us nothing except the constancy of inconstancy. There is no possible location within empirical history from which one could view the entire course of the history of the human race. One cannot find any law by empirical means that would allow us to determine the overall direction of history in the endless alternation of progress and regression. That is why it seems impossible a priori to recognize the general course of the process in which our own thinking is located. Unless one constantly awaits the day of judgement[91] or indulges in sanguine hopes,[92] one will be driven by empiricism to the hypothesis that

> all the interplay of members of our species on earth ought merely to be regarded as a farce. And in the eyes of reason, this cannot give any higher a value to mankind than to the other animal species, whose interaction takes place at less cost and without any conscious understanding.[93]

Kant thus shows that empiricism must lead to the same sceptical conclusions in the field of the science of history as in the field of natural science. Just as they had woken him from his 'dogmatic slumber'[94] in the latter case, they necessitate a transition to transcendental philosophy here too. He therefore continues:

> Perhaps it is because we have chosen the wrong point of view from which to contemplate the course of human affairs that the latter seems so absurd to us. The planets, as seen from the earth, sometimes move backward, sometimes forward, and at other times remain motionless. But seen from the sun – the point of view of reason – they continually follow their regular paths as in the Copernican hypothesis.[95]

It is impossible to read these lines without recalling the preface to the second edition of the *Critique of Pure Reason*, in which Kant describes the revolution in the way of thinking he has brought about as a 'Copernican turn'. This clearly shows the problem in his philosophy of history: an a priori cognition of history

---

[90] Ibid., p. 180.

[91] Ibid., p. 179.

[92] Ibid.

[93] Ibid., p. 180.

[94] Immanuel Kant, *Prolegomena to Any Future Metaphysics, Revised Edition*, ed. and trans. Gary Hatfield (Cambridge University Press, 2004), p. 10.

[95] Kant, 'The Contest of Faculties', p. 180.

is possible if we are able to see 'from the sun', which is exclusively 'the point of view of reason'. Only in this way can one conceive of arriving at an answer to the question, 'Is the human race continually improving?' that is valid because it rests on a law that has been cognized a priori.

'How can we obtain such knowledge? We can obtain it as a prophetic historical narrative of things to come by depicting those events whose a priori possibility suggests that they will in fact happen.'[96] One could hardly formulate what Kant demands of 'prophetic' history more pointedly. In order to interpret this demand correctly one must understand that the term 'events' [*Begebenheiten*] refers not to empirical facts, but rather to a process of improvement in accordance with certain laws – that is, not the *apparent* history but the *intelligible* history of the human species. It is permissible to speak of an 'intelligible history' at this point, as Kant uses the phrase 'history of pure reason' (*Critique of Pure Reason*, B880); but we must expand on this term. We have seen that 'In man [...] these natural capacities which are directed towards the use of his reason are such that they could be fully developed only in the species, but not in the individual.'[97] So there must be a development of the *faculty* of reason as such. Although the principles of reason are timeless and immutable, reason represents them as tasks that humans, both as a species and as individuals, can only approach in an infinite process. This is the process of the *existence* of reason as it realizes itself in the development of the species. That what the species approaches in this process is nonetheless *reason* is something of which we do not become conscious empirically, but only through the reflective power of judgement. The transcendental consciousness remains the same, but our consciousness *of* transcendental consciousness has a history. Only if there is such an intelligible history – later known as the 'history of spirit' – can one speak of progress. Kant himself does not refer to 'intelligible history' as history of spirit; in contradistinction to the 'natural history of humans', he calls it 'moral history' [*Sittengeschichte*].[98] The meaning of this term is defined by the title *The Metaphysics of Morals*; within this framework, as we have seen, it is assigned to the 'metaphysical first principles of the doctrine of right'. Hence 'moral history' is the process in which the moral principle is realized through the development of frameworks of right that correspond to it, and therefore enable the existence of reason in history.

After outlining the issue in this way, Kant offers a solution to the problem he has posed whose form is so simple that it initially seems incomprehensible through its very simplicity:

> But how is it possible to have history *a priori*? The answer is that it is possible if the prophet himself occasions and *produces* the events he predicts.[99]

---

[96] Ibid, p. 177.

[97] Kant, 'Idea for a Universal History with a Cosmopolitan Purpose', p. 42.

[98] Kant, 'The Contest of Faculties', p. 178; the published translation uses the inaccurate phrase 'history of civilization' (trans.).

[99] Ibid., p. 177.

In order to understand the context from which this answer arises, we must recall the 'Third Proposition' from Kant's first outline of his philosophy of history that we discussed above:

> Nature has willed that man should produce entirely by his own initiative everything which goes beyond the mechanical ordering of his animal existence, and that he should not partake of any other happiness or perfection than that which he has procured for himself without instinct and by his own reason.[100]

The answer to the question of how it is possible to have history a priori thus rests – as a systematic approach demands – precisely on the basic principle from Kant's transcendental deduction of history, in which the two parts of this deduction – the process according to natural principles and the process according to the principle of freedom – are intertwined. It is a fact of nature based on the *lex parsimoniae* that humans must produce their history themselves, that is, the spontaneity of their faculty of reason influences the shaping of history. But if reason is put into action at all over the course of history, it is inevitable that principles of reason, however sporadically at first, will affect the shaping of historical conditions. To the extent that they do so, one can say that 'the prophet', i.e. pure reason, 'himself occasions and *produces* the events he predicts'.

Such a sudden, chance appearance of reason in the public sphere of society would not, however, be sufficient to ensure the progression of history according to principles of reason. Kant's theory, as we have seen, is built in such a way that the guarantee of progress lies not in reason, but in a natural process that pushes the human species – even against its will – in the direction that would be demanded by reason if humans could act freely. On the other hand, the natural process alone is not sufficient to ensure a truly rational shaping of public conditions on this planet. For the mechanism of nature must be such that it would even outwit a 'nation of devils'. Reason, by contrast, is only possible out of freedom, and only what is produced freely is in keeping with reason. The natural process, as we have seen, is only a *conditio sine qua non* of history. But if reason is to occasion and produce its history '*itself*', that is, from spontaneity, it must be shown how, in the course of history, the reason that is latent in the human species as a capacity might not be put into action merely by chance, but could rather come into its own in such a way that it is empowered to itself through self-knowledge. This occurs, as we have seen, through the singular act of the 'revolution in the way of thinking' undertaken by Kant in the *Critique of Pure Reason*, which he also invokes through his reference to Copernicus in Chapter 4 of the second part of 'The Contest of Faculties'. All that remains is for us to examine how the Copernican turn is characterized in Kant's philosophy of history.

---

[100] Kant, 'Idea for a Universal History with a Cosmopolitan Purpose', p. 43.

The Copernican turn is brought about by the realization that reason 'has insight only into what it itself produces according to its own design'.[101] One need only recall this principle to discover that it is identical to that of 'prophetic' history:

> But how is it possible to have history *a priori*? The answer is that it is possible if the prophet himself occasions and *produces* the events he predicts.[102]

The transcendental principle of the second part of Kant's philosophy of history (deduction from the principle of freedom) thus arises directly from the principle of transcendental philosophy as such, namely the principle of the self-knowledge of reason. Reason can only achieve existence in history if pure reason, empowered to its own freedom by its self-knowledge, produces the future history of humanity according to its own design. If it is to achieve this, it must bring the framework of international right into alignment with its transcendental principles. It must derive the maxims of its public action from the 'universal principle of right' and remain in agreement with the 'transcendental formula of public right'. The 'hidden article on perpetual peace' thus forms the keystone not only of Kant's philosophy of public right, but also his philosophy of history. Humanity's progress towards improvement is possible a priori if the following principle applies: 'The maxims of the philosophers on the conditions under which public peace is possible shall be consulted by states which are armed for war.'[103] This assertion, as now becomes apparent, supports the title of the text in the strictest, transcendental sense of the word: 'Perpetual Peace: A Philosophical *Sketch* [*Entwurf*] by Immanuel Kant'. Perpetual peace is the state that reason is meant to bring about in history according to its own design [*Entwurf*]. Thus reason itself constructs the law whereby history can be grasped a priori as the progress of humanity.

The claim about reason that 'the range of its projects [*Entwürfe*] is unbounded'[104] is thus indeed the highest principle, in which the deduction of history as a natural process converges with the deduction of history from the principle of freedom. But Kant's 'critique of historical reason' can only be understood and justified as one part of the 'teleologia rationis humanae' whose architectonic he conceived in the *Critique of Pure Reason*. The outline we have given is therefore still incomplete; to complete our 'great project [*Entwurf*]' we would have to proceed 'from what experience makes immediately available to us, the *doctrine of the soul*, to the doctrine of the world and from there all the way to the cognition of *God*.'[105] Therefore, Kant's philosophy of history, along with the transcendental grounding of international right contained in it, is 'metaphysics' in a new, transcendental sense of this term. Within Kant's metaphysics of finite reason, however, this philosophy of history has a privileged status. Kant precisely defined its systematic locus: the final chapter of the *Critique of Pure Reason* bears the heading 'The History of Pure Reason', and below it there

---

[101] Kant, *Critique of Pure Reason*, p. 109.

[102] Kant, 'The Contest of Faculties', p. 177.

[103] Kant, 'Perpetual Peace', p. 115.

[104] Kant, 'Idea for a Universal History with a Cosmopolitan Purpose', p. 42.

[105] Kant, *Critique of Pure Reason*, p. 408n.

is the following statement: 'This title stands here only to designate a place that is left open in the system and must be filled in the future.'[106] We have arrived at this conclusion: Kant's transcendental grounding of international right is the doctrine that fills this gap.[107] It is the pinnacle of his philosophy of history.

[106] Ibid., p. 702.

[107] What Kant actually presents in this chapter is evident in the second sentence: 'I will content myself with casting a cursory glance from a merely transcendental point of view, namely that of the nature of pure reason, on the whole of its labours hitherto'. As the preface to the second edition shows, he views the previous history of philosophy as a mere prehistory. But it is a countercheck on his exposition of the 'nature of pure reason' that all previous attempts are compatible with its typology. The intention of the whole chapter is only made explicit at the end: 'The critical path alone is still open' (p. 704). The 'history of pure reason' in the strict sense of the word only begins with the opening of the critical path. Kant's philosophy of history serves to show how the 'history of pure reason' would be possible as the history of its existence a priori.

Chapter 6
The Intellectual-Historical Background of the Doctrine of Human Right (1957/1980)

In our historical and political consciousness, democracy and the defence of human rights are inextricably connected.[1] When we defend our basic *rights* we simultaneously defend a complex of basic *beliefs* that developed in the Romano-Germanic nations of Christian Europe. Their theoretical basis is only rarely examined, as political doctrines, national constitutions and finally legal dogmatics have interpreted these historically-developed norms as universal human rights; they did this in order to ensure their inviolability. But we cannot overlook the fact that the doctrine of human rights was also shaped by philosophical and political positions that are no longer tenable today. There is a danger that the baggage of these doctrines can distort the moral and political motives that require us – especially in the current world situation – to support material parts of the catalogues of human rights, and can expose them to ideological abuse. The ethics symbolized for us by the term 'human rights' is no longer identical to the doctrine of human rights. It is therefore time for an intellectual-historical re-evaluation that will clarify the implications of this doctrine. The following reflections offer a brief outline of certain perspectives that should be taken into account.

The method used here to link historical and systematic lines of thought may seem strange, but it corresponds to the way in which warp and weft are interwoven in the fabric of history itself. Historically and situationally specific tendencies draw on systematic 'patterns' in order to articulate themselves comprehensibly.[2] Because of its immanent logic, the systematics of these patterns later reasserts itself behind

[1] This text was originally published as: "Zum geistesgeschichtlichen Hintergrund der Lehre von den Menschenrechten". In: J. Delbrück/K. Ipsen/D. Rauschning (eds.), Recht im Dienst des Friedens: Festschrift für Eberhard Menzel zum 65. Geburtstag. Berlin: Duncker and Humblot 1975, pp. 289–305. Also in: Hier und Jetzt Vol. I, pp. 116–136. The permission to include this text here was granted by Klett-Cotta.

[2] The culture-morphological concept of 'patterns of culture' was introduced by Ruth Benedict in order to describe the structures of primitive cultures. I use the term in a somewhat altered sense to show that in highly developed cultures, thought structures that are normally viewed only from an

E. Rudolph and J. Picht (eds.), *Georg Picht: A Pioneer in Philosophy, Politics and the Arts*, Pioneers in Arts, Humanities, Science, Engineering, Practice 19,
https://doi.org/10.1007/978-3-030-31790-4_6

the backs of those involved, so to speak. If one is to understand the meaning and function of human rights in the history of the last three centuries, one must make this interplay of historical tendencies and systematic models transparent. The continuity of European history rests on the constancy of intellectual structures that only rarely enter the consciousness of its actors, remaining concealed far beneath the surface of the historical process. If one attempts to bring these structures to light, one arrives at a picture of history that is comparable to an X-ray image. The fact that such an image is not identical to a representation of the surface, and can even contradict it at times, is the reason why X-ray images are necessary.

I

The great Roman jurist Gaius begins his *Institutes* with the following words:

> All peoples who are ruled by laws and customs partly make use of their own laws, and partly have recourse to those which are common to all men; for what every people establishes as law for itself is peculiar to itself, and is called the Civil Law, as being that peculiar to the State; and what natural reason establishes among all men and is observed by all peoples alike, is called the Law of Nations, as being the law which all nations employ. Therefore the Roman people partly make use of their own law, and partly avail themselves of that common to all men [...].[3]

Here the *ius gentium*, the law of peoples, is understood not as a *ius inter nationes* but a *commune omnium hominum ius*, as a universal human law, that is, a law that applies to all peoples simply because they are humans. The particular legal systems of individual states are superimposed on this fundamental system, as it were. Montesquieu's adoption of this perspective demonstrates how enduring it was:

> Law in general is human reason in so far as it governs all the peoples of the earth; and the political and civil laws of each nation should be only the particular cases to which human reason is applied.[4]

For both Gaius and Montesquieu, the shared law for all humans is based on the fact that, as humans, they have a common nature. This nature is defined by reason. The conception of a universal human right did not grow on the soil of Roman law, which had very different notions about the *ius gentium*. It has no legal origins at all, but is based on a reception of Stoic philosophy that, from Cicero onwards, also exerted a strong influence on Roman legal thought.[5]

When the Stoics taught that the logos defines the nature of humans, they had something quite different in mind from what we call 'reason' today. The word 'logos'

---

intellectual-historical perspective are, in their function of structuring societal behaviour, similar to Ruth Benedict's 'patterns of culture'.

[3] *The Institutes of Gaius*, http://www.thelatinlibrary.com/law/gaius1.html.

[4] Montesquieu, *The Spirit of the Laws*, ed. and trans. Anne M. Cohler, Basia Carolyn Miller and Harold Samuel Stone (Cambridge University Press, 1989), p. 8.

[5] This is not the place to show that the concept of *ius gentium*, which first appears in Cicero (*De officiis*, 3, 69), comes not from Aristotle's concept of natural law but from the Stoics (see Max Kaser, *Das römische Privatrecht* [Munich: C. H. Beck, 1975], III, 3, 1, p. 182). The decisive element is the concept of *ratio* presupposed by Gaius.

does not primarily refer to human reason; rather, it refers to the nature of God, in so far as this nature manifests itself in the fact that God as logos is the ground of the world's unity. As a grounding unity, the divine logos gives the world its law: it is *nomos*. Materially, the divine logos manifests itself in the world as the primal fire from which all elements emerge. The special position of humans in the cosmos stems from the fact that they carry a spark of this primal fire in their souls. That is why they carry in them the *capacity* to join with the divine logos and recognize the law of the cosmos. This is the Stoic reinterpretation of the Aristotelian doctrine that the human being is the being that has the logos. The so-called 'reason' of the human being is not its own reason, but rather the divine logos that manifests itself in it.

From the perspective of this metaphysical doctrine we can explain the basic formula of Stoic ethics: *homologoumenōs tē physei zēn*. The Romans translated it thus: *secundum naturam vivere* – 'live according to nature'. With this simplifying translation, however, the philosophical content of the idea is lost:

1. The word *homologoumenōs* refers to the metaphysical doctrine that the logos of humans is identical (*homo*) to the divine logos. This is the basis for the harmony between the human logos and the logos of nature.
2. On the basis of this doctrine, *physis* means both the nature of humans and nature as a whole; for it is precisely the nature of the logos that, by virtue of its divine origin, it keeps the nature of humans in harmony with nature as a whole. Both obey the same nomos, which consists for human reason in the capacity to grasp its unity with the divine logos within itself.

This is the foundation on which the Stoics developed a metaphysical doctrine of law that would determine the entire later view of so-called human rights. In so far as humans obey the logos they are, according to Stoic teaching, citizens of a *civitas* that equals no earthly community: the *civitas* of gods and humans (according to the Stoics, all the gods in popular religions are different manifestations of the One Logos). Thanks to their reason they are 'cosmopolitans' (i.e. citizens of the universe) and accordingly have an originary civil right in a stateless space. Hence the background to the distinction between *ius gentium*, which is based purely on reason, and the *ius civile* of different peoples is the distinction between the metaphysical *civitas deorum ac hominum* and the historical *civitates*. Augustine famously adopts this stoic doctrine in the distinction between *civitas dei* and *civitas terrena*; he wrote out the corresponding sections from Cicero's *De re publica* at length.[6] It was not inconsistent that the secular thought of the Modern Age discarded Augustine's transformation of the Stoic two kingdoms doctrine once more in the seventeenth and eighteenth centuries. The Modern Age did not establish any reformulation of the doctrine of human rights, however, but returned to the pre-Christian basis of Stoic philosophy. This was in keeping with the spirit of the age. Between the sixteenth and eighteenth

[6] On Augustine's reinterpretation of Stoic teachings, see Ulrich Duchrow's book on the two kingdoms doctrine, *Christenheit und Weltverantwortung: Traditionsgeschichte und systematische Struktur der Zweireichelehre* (Stuttgart: Klett-Cotta, 1983), pp. 247ff.

centuries, as Dilthey has shown,[7] a reception of the Stoics took place that was no less historically significant than the reception of Aristotle in the thirteenth century. It is no great surprise, then, that the doctrine of human rights also makes use of the traditional Stoic formulas; Stoic philosophy supplies the 'pattern' for the defence of ancient rights to freedom that had arisen in very different circumstances. But this simultaneously means a reinterpretation of these rights to freedom; they are universalized with the help of Stoic formulas. The universality claimed for human rights now only exists as a postulate, however; it is no longer grounded in a metaphysical conception of the world. For human reason is no longer understood as a direct manifestation of all reason that governs the whole of nature. The 'nature' of the human being is still derived from its faculty of reason, to be sure, but the doctrine that human reason is identical to the basic law of the cosmos can no longer be maintained. The correspondence of the 'laws' of reason to the laws of the rest of nature is undone. As a result, the concept of reason itself inexorably loses its absolute validity. The historical qualification of the European concept of reason, however, sends tremors through all the doctrines founded on it – including the doctrine of human rights.

II

Once the fabric of metaphysical equations on which the Stoic doctrine of human rights rests had been broken apart, the following problems had to ensure:

1. Reason is no longer a universal law of nature (although Kant attempted to salvage this connection in the categorical imperative). Theoretical reason became a mere faculty of judging and concluding, and therefore has no content of its own from which one might derive a universal human right; practical reason is predicated on a concept of freedom of such manifestly European-Christian origins that its universal validity becomes a mere postulate.[8] This eliminates the possibility of interpreting the *ius gentium* as a fundamental human right on which the positive legal frameworks of the individual states are simply superimposed. On the contrary: from the end of the seventeenth century, the growing knowledge of foreign cultures, religions and peoples in Europe led to a greater insight into the relativity of all mores, customs, forms of faith, morals and religions, including European norms. This resulted in the 'crisis of the European mind' described by Paul Hazard.[9] A now global empiricism refuted the opinion that the same natural law must apply among all peoples purely on the basis of human reason. This did away with the equation 'universal human rights = basic rights', which now became merely an abstract demand that could be fleshed out at will according to the respective worldview or tradition. This gives us cause to remember once

[7] See Wilhelm Dilthey, *Weltanschauung und Analyse des Menschen seit Renaissance und Reformation* (Göttingen: Vandenhoeck & Ruprecht, 1977).

[8] See Georg Picht, 'Der Sinn der Unterscheidung von Theorie und Praxis in der Philosophie der Neuzeit', in *Wahrheit, Vernunft, Verantwortung*, pp. 135ff.

[9] Paul Hazard, The Crisis of the European Mind: 1680–1715, trans. J. Lewis May (New York Review of Books, 2013).

again that the origins of this equation are metaphysical, not legal. And as far as I can tell, jurists have never used the logos of the universe to interpret basic rights.

2. Metaphysics does, however, re-enter the grounding of human rights through a back door. This back door bears the inscription 'person'.[10] To quote Jellinek:

> The recognition of the individual as a person is the foundation of all legal relationships. Through this recognition, however, the individual becomes a member of the people in its subjective quality. [...] The recognition as a person and member of the state is the basis for all claims subject to public law, which are consequently divided into those granted by the state to all people within its area and those reserved for people who belong to it permanently as citizens.[11]

This presupposes the following:

(a) All people are 'persons' by virtue of the simple fact that they are human beings;
(b) As such, they have a 'subjective public right of the individual' that must be recognized by any state order.

In this theory, the Stoic schema of fundamental rights that are common to all human beings on the basis of their human nature, and on which the positive legal systems of individual states are merely superimposed, comes to the fore again. The concept of 'person' presupposes the modern philosophical doctrine that reason is based on freedom. The principle that the human being is a being gifted with reason now implies the demand that its freedom must be recognized. This is viewed as 'the foundation of all legal relationships' and must therefore be respected by all states. We are here located within the Stoic schema of Gaius, except that the universal logos has been replaced by the likewise metaphysical nature of freedom.[12]

---

[10] The connection between the terms *persona* and *humanitas* was adopted by Cicero from the ethics of the middle Stoic Panaetius (see my unpublished dissertation 'Die Grundlagen der Ethik des Panaitios, Freiburg 1943), and subsequently became one of the most influential elements within the European humanistic tradition. Independently of that, Roman legal theory also used the word *persona* in the sense of 'mask' or 'role' in order to approach the concept of 'legal capacity'. The unbridgeable gap between these two understandings of *persona* shows itself in the fact that as a legal term, it is also used with regard to corporations and foundations (see Kaser, *Das römische Privatrecht*, II, pp. 75ff.). These two meanings are merged in the modern usage of 'person' and interpreted in different ways according to the respectively dominant philosophical direction. In the case of Jellinek, the background to his concept of 'person' is a neo-Kantian interpretation of Kant's philosophy of right.

[11] Georg Jellinek, *Allgemeine Staatslehre* (Berlin and Zurich: Gehlen, 1966), p. 419.

[12] Jellinek derives the doctrine of subjective public right from the dualism of church and state, and the Christian doctrine of the freedom of religious conscience, which subsequently merged with 'the never entirely abandoned Old Germanic view of the priority of individual right' (p. 411). What is problematic about this thesis is not only that it interprets Old Germanic 'freedoms' as modern 'individual rights'; Jellinek has also been criticized for asserting that the origin of human rights was freedom of religion (see Martin Kriele, *Einführung in die Staatslehre* [Reinbek: Rowohlt, 1975], pp. 151ff.). But this objection strikes me as missing the heart of the matter. For what matters is not

3. If one does not deduce the 'basic rights' systematically from the nature of humans but derives them historically, they are the 'freedoms', privileges or immunities that were retrieved for the constitutional state from pre-state times. Carl Schmitt observes:

> The Magna Carta of 1215 [...], the Habeas Corpus Act of 1679 [...] and the Bill of Rights of 1688 [...] are often termed the first declaration of basic rights. They are, in fact, contractual or statutory regulations of the rights of English barons or citizens, which in the course of a gradual development certainly assumed the character of modern principles, but they do not correspond to the original meaning of basic rights. The history of the basic rights, more precisely, first begins with the declarations the American states established in the eighteenth century to justify their independence from England. As Ranke expressed it, this is really the beginning of the democratic, more accurately, of the liberal age and of the modern bourgeois, liberty-based Rechtsstaat, although these American declarations designate themselves still as 'Bills of Rights' in connection with the English tradition. [...] The essential basic rights of these declarations are freedom, private property, security, right to resistance, freedom of conscience, and religious freedom. The securing of these rights constituted the purpose of the state.[13]

The intention of this text is to put the basic rights into question through historical examination. If it can be shown that the content of the catalogues of human rights emerged from the contractual or statutory regulations of English barons and citizens, they can no longer be claimed as universal human rights that, because of the rational nature of humans, must be respected by all peoples in all states. But this also applies to the declarations of the American state from the seventeenth century, for their catalogues follow on from the English legal tradition. One must therefore say that Carl Schmitt did not succeed in precisely locating the point of the shift to the bourgeois-democratic era.

The new and revolutionary element lies in an idea taken not from the English legal tradition, but from French philosophy: the idea that the basic rights of a particular national constitution should take on the universal application of general human rights, and that the privileged status of certain classes within a particular nation should be extended to the whole of humanity. The basic rights are taken out of the framework of the English constitution, absolutized and transferred to the ancient Stoic schema

---

how much significance freedom of religion is granted in the catalogue of basic rights; rather, the question is what understanding of humans can ground the totality of these righs if one postulates that they should be considered universal human rights. With the transition from the basic rights of a particular constitution to the postulate of universal human rights, the question of the source of legitimacy for so universal a postulate takes on a new function. A universal doctrine of human rights cannot be derived from particular developments in constitutional history. That is why, *pace* Kriele, jurists are forced to let philosophers have a say. Hence Jellinek's emphasis on natural law as a ground of human rights cannot be disputed. Then, however, we must counter Jellinek by pointing out that, in dogmatic terms, the 'dualism' of church and state took shape through the two kingdoms doctrine, which presupposes Augustine's reception of the Stoics, that is, the Stoic 'pattern' of Gaius. It is therefore permissible for our 'X-ray image' to see a connection between the modern doctrine of subjective public law and the Stoic schema.

[13] Carl Schmitt, *Constitutional Theory*, trans. Jeffrey Seitzer (Durham and London: Duke University Press, 2008), p. 197.

of Gaius. Thus they define – and here one must agree with Schmitt – a determinate, historically new and never before actualized form of the state: the bourgeois constitutional state. This constitutional state is bourgeois because the catalogues of basic rights, as is easily shown, summarize the central demands of eighteenth-century bourgeois ideology. The word 'ideology' is appropriate here; for in the United States too, it immediately became clear that there can be repercussions if, in a historically unique state system, one approves the achievements of the higher classes as universal human rights without examining the historical circumstances and natural conditions in which humans live. The Native Americans and the slaves did not benefit from 'human rights'; in fact, these rights served for a long time to secure the privileges of the white population of the United States. That people have meanwhile begun to take them seriously is leading to conflicts that will not be resolved in the near future.

III

The deliberations in sections I and II were necessary in order to make visible the philosophical and socio-political preconditions for the individual rights identified in the catalogue of basic rights. In philosophical terms, the individual rights rest on an anthropology that views every human being as a 'person', i.e. an individual who is capable of reason and therefore granted a claim to freedom by nature. This idea corresponds to the legal status demanded for every *civis* by bourgeois society. The specifically bourgeois character of the classical basic rights is clear from the central position occupied by property in the catalogue of basic rights. Under the mask of 'individual rights', the basic rights are bourgeois class rights. This explains the later antinomy between individual and social rights.

To understand this antinomy, we must first take a closer look at the change in the understanding of rights that took place in the background to the doctrine of human rights once they were interpreted in a bourgeois sense as individual rights. Gaius speaks of the *commune omnium hominum ius* to refer to the totality of *laws* that all people *follow*, in so far as they make use of reason. The liberal catalogue of basic rights, on the other hand, encompasses the totality of *claims* to protection of civil liberties *made* by individuals based on their quality as persons. The addressee of these claims is the state, of which it is also demanded that it approve by means of its constitution the claims made of it. The relationship to the state shows that the liberal basic rights have a real function that is not mentioned in the doctrine of basic rights: they guarantee the system of freedoms that is necessary in order to establish an economic order based on the principles of liberal economics. This economics teaches that the interests of all are served best if every individual acts according to their individual interests. According to this theory, it is therefore best for the common good if the state regulates as little as possible. This leads to an expansion and reinterpretation of freedom, which, according to the 'liberal' conception of law, is meant to be protected by human rights. The old union of Christian freedom of conscience with Germanic 'freedoms' and Roman civil rights is now augmented by the 'free' play of interests in an economy of 'free enterprise'. The reinterpretation of the highest moral obligations as rights to resistance under public law, and of freedom of conscience as the 'right' to unlimited representation of personal interests, can only be understood against this background.

But we must not overlook the fact that, as a result of the interpretation undergone by the concept of freedom in the economic theory of liberalism, it paradoxically became possible to connect it once more to natural law. The theorists of liberalism famously believed they had identified the laws of nature obeyed by the physics of economic and social forces. If the basic rights guarantee the state-free space in which the free play of interests cannot be prevented, this restores the harmony with nature that was disrupted by the wilful statutes of the old feudal system. Thus, in a peculiar way, the Stoic demand of *secundum naturam vivere* is reinstated; the logos is replaced by the 'invisible hand'. What seems questionable to us – the reinterpretation of the metaphysical freedom of person as the free play of egotistical interests – is justified by the fact that the concept of interest mediates between moral freedom and nature.[14] As long as the consequences of capitalist economy were not yet visible, it was difficult to detect the devilries concealed in this ambiguity.

In the eighteenth century, the economic development enabled by the progressive emancipation of the bourgeoisie and later theoretically explained by Adam Smith led to the dissociation of state and society first theorized by Montesquieu. Now it was not only the individual but also society that was fighting for its independence from the state. Admittedly, it was soon split up into a multitude of quarrelling classes. This process gave rise to the idea of augmenting the classical individual rights with 'social human rights'. One can now no longer speak of an ambiguous legal conception; human rights have openly been declared legally approved entitlements to benefits from the state. They indicate the sum of social demands that the welfare state is meant to fulfil.[15] In their content, the social rights have an ethical and political legitimacy that is at least as well-grounded as that of the individual rights. What is problematic, however, is their subsumption under the overall category of 'human rights'. It can be justified neither anthropologically nor historically, but is explained by the need to give the new legal claims the dignity of a title in which the metaphysical and religious reminiscences of two thousand years of European history are still alive.

But the dissociation of state and society, in turn, follows the 'pattern' of the old Stoic schema. The state-free space established in the name of the person through the basic rights was initially understood as the space of morality in contrast to legality. This is a diluted memory of the Stoic two kingdoms doctrine, which was from the start the foundation for the idea of a *commune omnium hominum ius*. Because morality can also be interpreted in a Christian manner, Christians have always had the possibility to project ideas and motives into the state-free space of morality that originated from Augustine's reinterpretation of the Stoic two kingdoms doctrine. Because the state-free space of moral freedom was simultaneously used for the free play of economic interests owing to the economic theory of liberalism, however, the

---

[14] Concerning the influence of the economic theory of liberalism on Kant's philosophy of right, see 'Kant's Transcendental Grounding of International Right'.

[15] Regarding these problems, see Ernst Friesenhahn, *Der Wechsel des Grundrechtsverständnisses – Verhandlungen des fünfzigsten deutschen Juristentages* (Hamburg: G. Jansen, 1974), II G, pp. 14ff.

consequence of the moral-philosophical secularization of the two kingdoms doctrine was the creation of an empty space for social emancipation beyond the state. This put paid once and for all to the classical concept of reason, for reason now appears in three antinomic forms. They are:

1. Reason of the subject of individual rights
2. Reason of the *volonté générale* of collectives that immediately discover their class oppositions and thus destroy the universality claimed by the term *volonté générale*
3. The reason [*Raison*] of the state that is simultaneously meant to guarantee, restrict and fulfil the claims asserted against it.

Because the state faces both individuals and subjects of collective claims (associations etc.) in a state-free space, it can no longer act as the subject of the common good, for the subjects of social interests have dissociated themselves from it. Therefore, in industrial societies, it increasingly develops its purely instrumental functions. It changes into a 'state apparatus'. This aligns with the interests of those forces that see it merely as an instrument of class struggle.

The objective fragmentation of reason among conflicting subjects has inescapable consequences for the understanding of human rights. From Gaius to Montesquieu, the commonality of reason formed the basis for the idea of universal law among humans. For Kant, human law, as a law of reason, still encompasses the sum of duties that are binding for humans on account of their rational nature. Owing to the fragmentation of reason into opposing subjects, the state can no longer appear as the subject of the common good in keeping with the classical theory of the state; on the other hand, the subjects of individual or social rights no longer understand these rights primarily as obligations, but as rights *to* something. The far-reaching reinterpretation of the concept of rights, which is reflected especially clearly in the political or indeed demagogical misuse of human rights slogans, is the result of the great upheaval in all economic, social and political orders caused by the technological-industrial revolution. Established political structures are undermined and no longer offer any support for legal conceptions. The classical state order finds itself in a process of dissolution and cannot be reinstated, for it stands in an irresolvable contradiction to the economic and political conditions of the technological world. New foundations for legal legitimacy have not yet emerged. The fascination emanating from the doctrine of human rights today rests on the fact that it seems to fill what is generally perceived as a vacuum. Where law fails, it allows a shift to ethics, and where violence is done to ethics, it permits an invocation of law. But this play of consciousness, as the cruel realities of our world prove, is illusory. If we understand so-called 'human rights' as basic rights, they stand or fall with the order of the state whose constitution guarantees them; if the entire state order is undermined, this must also endanger our basic rights. If, on the other hand, we understand them as norms that stand above and outside the state and are valid for the whole of humanity, regardless of differences between religions, cultures and political systems, one would have to provide a theoretical grounding of this universal validity. This would only be possible on the basis of an anthropology accepted by all the peoples of the world. Having seen that the anthropology on which the traditional doctrine of human rights is predicated has been irrevocably shattered even in its European homeland, we can only view the utopia of a global order of human rights as an empty delusion.[16]

IV

Human rights are an attempt to transpose convictions that belong to a pre- and extralegal sphere into the sphere of law. As 'basic rights' they claim a position that takes precedence over all other legal principles. By declaring them 'human rights', one asserts that they possess absolute validity and must be respected by all nations at all times. If one attempts to analyse the thought figure of human rights in terms of its systematic scope, one must examine the following:

1. whether and within which boundaries convictions that belong to a religious or moral sphere can be transposed into positive law;
2. what source of legitimacy there is to assert that certain positive norms possess unconditional validity for the whole of humanity;
3. what premises underlie the conviction that absolute norms of human action and timeless, immutable qualities of human nature can be cognized.

Regarding 1: in the background to the doctrine of human rights there is a Greek legal theory we have not yet examined. The legal philosophy of the fifth century BCE already distinguished, as every reader of *Antigone* knows, between 'unwritten' laws (*agraphoi nomoi*) and written laws. The unwritten laws come from the gods, and therefore apply to all humans. The written laws, on the other hand, are the work of humans. They are an instrument of politics and can be changed at will. In Greek thought, this distinction is heightened into an opposition because humans, on account of their mortal nature, are subject to the power of deception and cannot cognize the unadulterated truth of the divine laws, and because whatever part of them they do cognize cannot be transferred to the relativity of positive statutes. Antigone and Creon both obey an inalienable necessity: both of them are at once right and not right. In right, a contradiction manifests itself that is irresolvable and therefore tragic.

The modern doctrine of human rights also reflects the spirit of the eighteenth century in the fact that no distinction, much less an opposition is acknowledged between what is available to humans and what is decreed for them by the gods. Rationalism believed in the autonomy of human reason. It was convinced that reason is capable of subjecting the totality of human living conditions to its own legislation, i.e. its sovereign self-rule. Human rights are a symbol of this modern faith in reason. They transform the unwritten laws into written statutes. The French were consistent enough to see that the state would then also have to be changed in such a way that the tragic conflict between Creon and Antigone could no longer ensue. §2 of the Declaration of the Rights of Man and of the Citizen of 16 August 1789 presents the following definition: 'The goal of any political association is the conservation of the

[16] I need hardly emphasize that this statement is not intended to cast doubt on the moral, political and legal legitimacy of the pragmatic struggle for an improvement of the situation of individuals or social groups in different political systems, especially in the area of 'basic judicial rights'.

natural and imprescriptible rights of man.' As no state in world history before the United States could be said to have served the purpose of preserving human rights, this definition denies all pre-revolutionary political formations the character of the 'state'. This corresponds to the spirit of the French revolution. It shows very precisely how the new revolutionary state is founded through the codification of 'unwritten laws'.

After the experiences Europe and the world have gained since then with the destructive consequences of the idea of autonomous reason, there is reason to ask whether the Greek doctrine of the unavailability of pre- and extralegal preconditions for all legal systems does not contain a deeper knowledge of human nature and the conditions of humanity than the anthropology of the Modern Age that found expression in the catalogues of human rights. But one can certainly learn from the French declaration that, thanks to the transposition of unwritten into written law, human rights were simultaneously linked inextricably to the human product that was the modern constitutional state. They broke out of the sphere of religion and ethics and were built into the structures of a rational state framework. One can now no longer speak of inalienable human rights in any ethical or religious context outside of rationally planned institutions and without the precision of legal language, for the constitution made the protection of these rights dependent on the state's capacity to function. Having been transformed into positive statutes, human rights can now only be defined in the context of the overall constitution of the state and society. They are now identified with a particular historical form of state constitution; but this is precisely why they have lost their claim to absolute validity.

Regarding 2: in both form and content, the thought figure of human rights, in spite of its philosophical and religious prehistory, is the symbol of a revolutionary innovation. None of the advanced civilizations respected the totality of the rights contained in the various human rights catalogues. None of them recognized the fundamental principle that stands at the top of the French declaration: 'Men are born and remain free and equal in rights.' This has not changed since then. Ludwig Raiser shows impressively in his essay 'Human Rights in a Divided World'[17] why even the United Nations could not reach a consensus on the details of human rights after its declaration of 1948. The emphasis on personal autonomy and freedom also in relation to the state, which had entered the UN declaration from the European-American tradition, was rejected by the socialist states led by the Soviet Union, and both factions in turn contradicted the demands that the Third World sought to assert under the name of 'human rights'. Ideas about the content and function of human rights diverge so massively that the notion of universal norms that are valid for the whole of humanity proves a hollow mould, whose declamatory value lies in the fact that it can be filled with any chosen content. One can, with a hint of cynicism, hold the opinion that precisely therein lies the demagogical utility of a formula which has still not lost the glamour of the reminiscences attached to it. But anyone who seriously believes that the humanity of humans can be defined or protected by universal norms

[17] Ludwig Reiser, 'Menschenrechte in einer gespaltenen Welt', in *Evanglische Kommentare*, April 1975, pp. 99ff.

must not evade the enormous burden of argumentation they will have to carry. For what they are declaring about the 'nature' of humans is refuted by the whole of human history.[18]

In this context we should also remember that there are conditions of humanity which are not mentioned in any of the human rights catalogues. Antigone sacrificed her life for the right of the dead. She believed it to be an unwritten law of the gods that the humanity of the living is destroyed if they do not understand how to honour the right of the dead. This right does not appear in the concept of 'person'. But perhaps there is reason to doubt whether the rhetoric associated with the concept of human rights makes legitimate use of the word 'sacred'.

Regarding 3: human rights are viewed as inalienable, valid for all people at all times, i.e. as absolute norms. Anyone who seeks to declare an absolute norm at once claims an absolute knowledge.[19] Indeed, European metaphysics taught that human reason is capable of an absolute, timeless cognition. This doctrine relied philosophically on the ontological proof of God and theologically on the dogma of man as the image of God. If one calls into question the basic tenets of European metaphysics, one simultaneously unsettles the doctrine of human rights. For what should the identification of absolute norms be based on if no one dares assert the possibility of absolute knowledge any longer? The doctrine of human rights irrevocably implies – in form, not content – the absolute knowledge of European metaphysics. This is no longer familiar to us because metaphysics in Europe has entered the basic order of all spheres of human existence. In Christian theology it led to the conviction that the truth of faith can be expressed in timelessly valid dogmas. In the sphere of science it underpinned the way of thinking we tend to call 'systematic', for a 'system' is a context of justification for statements that can all be traced back to universal and timeless principles. In natural science metaphysics hides behind the belief in eternal natural laws. In ethics we find metaphysics behind the belief that all human action and behaviour is subject to timeless norms that can be identified and formulated much like the dogmas of theology or the principles of systematic science. How to implement this metaphysical approach in the legal realm is demonstrated both by the Catholic theory of natural law and the apriorism of Kant's philosophy of right.

The discussion of 'human rights in a divided world' cannot lead to any consensus if one evades the question of its source of legitimacy. Then, however, one will be taken back, with inescapable rigour, to fundamental problems of legal philosophy

[18] The assertion that universal norms are anthropologically impossible formulates the results of the investigation I contributed to the studies by the Committee for International Law of the Protestant Study Association. In this committee, Eberhard Menzel pointed tirelessly to the importance of pre-and extralegal preconditions for international law. My contribution was an attempt at least to answer a part of the questions posed by him.

[19] This applies – and it is not superfluous to underline this – not only to moral and social norms, but also to legal norms that are meant to have universal validity. For a legal norm must first be *recognized* in order to be formulated. Its validity must be *grounded*, especially if it contradicts the legal systems that apply in numerous states. The basis of legitimacy in this case can only consist in general principles that, if they are to be universal, must apply absolutely, which means that they can only be recognized through absolute knowledge.

that cannot be explained here. Suffice it to say that jurisprudence as well as – in a different context – moral philosophy and Christian theology face the suspicion of not taking human rights seriously if they fail to examine this thought figure's source of legitimacy and evade the problem of its philosophical justification. A brief overview of the human rights discourse gives the impression that the majority of partners are no longer aware of the jurisprudential implications of the doctrine of human rights. This leads not only to manifest weaknesses of legal, political, moral-philosophical and theological argumentation; it also has something to do with the material content of everything that is meant to be protected by human rights. Schiller's letters on 'aesthetic education' contain the prophetic words: 'In other parts of the world the humanity in the negro will be honoured, while in Europe it is disparaged in the thinker.'[20] Here, the word 'humanity' is not the collective term for the human species; it refers to the nature of the human being. In Schiller's time, it was still believed that the dignity of humans rested on their ability to think. For Europe, he prophesied the capitulation of thought that followed in the second half of the nineteenth century and caused all further ruin that befell this continent. There would be tragic consequences, not only for Europe, if the discourse on human rights became the setting in which the truth of Schiller's words would be demonstrated. That is the reason for writing this sceptical reflection on the formal thought figure of human rights in the service of their material content, which, for all the ambivalence of that doctrine, are defended whenever jurists, politicians or theologians in the world of today dare to stand up for 'human rights'.

Afterword (1980)

The intellectual-historical 'X-ray image' of the doctrine of human rights, originally published in the festschrift for Eberhard Menzel,[21] was written before President Carter declared his human rights policy. Whatever one thinks of this policy, it has certainly led to the formation of ideological fronts that make an unbiased study of the facts, theories and ways of thinking that clash in this field almost impossible. Whoever voices a critical opinion on the doctrine of human rights is suspected of advocating political systems that shy away neither from torture nor genocide, neither from suppression of freedom nor mockery of human rights.

It must be permitted, however, to ask whether our helplessness in the face of the wave of horror currently washing over the Earth is perhaps connected to the fact that our grounding of human rights is insufficient, our theoretical tools inadequate and our intellectual foundation brittle. The doctrine of human rights emerged from a great tradition; its venerable status and moral substance rest on the fact that it implies the universality of the issues of European metaphysics. For the consciousness of our civilization, however, metaphysics is a thing of the past; it has been forgotten, misjudged and disregarded. There are compelling reasons for the impossibility of

---

[20] Friedrich Schiller, *On the Aesthetic Education of Man*, trans. Keith Tribe (London: Penguin, 2016).

[21] *Recht im Dienst des Friedens. Festschrift für Eberhard Menzel zum 65. Geburtstag*, ed. J. Delbrück, K. Ipsen and D. Rauschning (Berlin: Duncker and Humblot, 1975).

returning to the metaphysical convictions of the last two and a half thousand years. Yet we cling to the theory of human rights; we try to defend it with the methods of legal positivism, taking recourse to moral convictions that we cannot legitimize either through our religion, a dominant philosophy or the social praxis of the industrialized nations. Religiously, intellectually and morally impoverished, we preserve human rights as an abstract doctrine that we – as proved by American policy – disavow through our actions. The essay presented here documents a feeling of shock at this situation. It seeks to show that the crisis of European metaphysics must inevitably lead to a corresponding crisis in the doctrine of human rights, and that the preservation of humanity will depend on whether we succeed in understanding its conditions better.

The international discussion triggered by President Carter's human rights policy led to numerous public statements that have also changed the landscape on the level of theory. The most important of these is the speech by Pope John Paul II to the United Nations. I will not discuss it here; my essay never claimed to do justice to the entire breadth of the problem in any case. But a few remarks on the political aspects may be helpful for purposes of clarification.

1. Individual human rights (the term is not identical to 'civil rights') only became effective as a structurally formative principle in those parliamentary democracies that adopted them in their constitutions as 'basic rights'. For all other states – whatever verbal statements they might make – a realization of individual basic rights could only be achieved through a revolutionary change in the existing political structures. The American government was well aware of this when President Carter initiated his human rights policy. I received the response from a relevant source at the State Department that American policy was determined not least by the need to aggressively counter the ideological warfare of the Russians.
2. Even in the democratic industrialized nations, social rights are not realized and cannot be claimed by legal means. A realization of social rights would revolutionize these states too.
3. It is a result of the Protestant traditions that flowed into the doctrine of human rights that religious freedom is usually interpreted as freedom of conscience, that is, as an individual right. This contradicts the perspective of other religions that know neither the concept of conscience nor our individual rights. Whoever demands religious freedom as a universal human right must know that they are espousing the recognition of ways of thinking that are irreconcilable with the doctrine of human rights. In a pluralistic system that respects grown religions and cultural structures, the view of religion crystallized in the doctrine of human rights appears as *one* alongside other equally valid ones. Whoever declares it universally true is seeking to convert the other religions to human rights, and thus demanding a new era of world mission.
4. The realization of human rights presupposes the survival of humanity. From this follows an ecological dimension to the problem that has not yet been brought up for discussion. Through social rights, the doctrine of human rights is tied inseparably to so-called 'basic needs'. Under the given conditions, therefore, it

> only becomes plausible if its economic implications are analysed and the corresponding steps taken with regard to the global economic order. It will probably then emerge that the entire context of human rights, including individual rights, is dependent on economic preconditions that cannot be fulfilled on a global scale within our ecosystem.

It follows from these reflections that the question of human rights takes us to the heart of a problem that cannot be overcome with the methods of legal positivism. Legalistic ways of thinking only access the surface of a global situation in which we must ask whether it is even possible to preserve the humanity of humans, or whether its demise is the prelude to global disasters that exceed our imagination.

In this situation, we should remember that codified human rights have a core that is inaccessible to both legislation and jurisdiction: human dignity is the axis of human rights. Human rights as such are merely an attempt to formulate certain conditions that should be given for human dignity to be possible. Where does human dignity as such originate?

Our current civilization is characterized by its failure to ask this question. I assert, without justifying it any further here, that humans have dignity when they recognize what is higher than them. They have dignity when they accept that there is that which is not available to them. Their human dignity is protected if they live in an order that respects such a stance and therefore also respects the unavailable essence of each individual person and the structures of their life. The translation of such general statements into ecological and social networks, political systems and legal frameworks is a task that clearly cannot be adequately solved using the concepts developed by our tradition in Europe and the United States. The rediscovery of humanity can only ensue in dimensions that are closed to our current mentality.

Chapter 7
Philosophy and Politics (1972)

It is dangerous to make politics a topic of philosophical thought; for there is an irresolvable opposition between politics and philosophy, as we have known since Socrates.[1] The utopia in which they might be reconciled is negated by the reality in which they encounter each other. The horizon of philosophy is what we call 'truth'; the medium of politics is the field of power.[2] The opposition between philosophy and politics is irresolvable because it stems from the distinction between power and truth, and because this distinction cannot be reduced to mere modifications of human consciousness, but rather comes from the phenomenality of phenomena themselves.

I have termed politics the 'field of power'. Some find this description offensive. As soon as one utters the word 'power', one immediately sets affects in motion that make it all but impossible to specify exactly what is meant by it. The nefariousness with which power is applied and used to threaten in the world of today stands in direct relation to the fact that in the society to which we belong, power is accompanied by similar taboos to those that apply to sexuality in bourgeois society. If one asks why this is the case, the question leads into ever deeper layers of collective consciousness. It is not only political realities, not only a collective malaise over the injustices of our social order that are suppressed; whenever the word 'power' is spoken, there are also reminiscences of the religious issues that modern consciousness represses because the process of the Enlightenment was unable to penetrate them. A closer investigation of the difference between the field of power and the horizon of truth would have to advance to the dark areas of myth and magic. The semantics of the term 'power' cannot be untangled as long as the relationship between faith and knowledge remains opaque.

[1] This text was first published as: "Philosophie und Politik". In: *Merkur* 26 (1972), pp. 617–623. Also in: *Hier und Jetzt* Vol. II, Stuttgart: Klett-Cotta 1981, pp. 13–22. The permission to republish the text here was granted by Klett-Cotta.

[2] An explanation of this term can be found in 'Ist eine philosophische Erkenntnis der politischen Gegenwart möglich?' in: *Hier und Jetzt* Vol. II, Stuttgart: Klett-Cotta 1981, pp. 229–332.

E. Rudolph and J. Picht (eds.), *Georg Picht: A Pioneer in Philosophy, Politics and the Arts*, Pioneers in Arts, Humanities, Science, Engineering, Practice 19,
https://doi.org/10.1007/978-3-030-31790-4_7

Because power is at once the golden calf of our society and the taboo area of its consciousness, the philosophical statement that 'the medium of politics is the field of power' immediately leads to the misunderstandings and conflicts that philosophy must expect if it dares to venture into this field. For in the field of power, the affects that would forbid us to recognize power for what it is themselves play with the means of power. The dynamics of power keeps their undulations in motion.

If we look more closely at the catchwords with which our society (or at least strong classes and groups in that society) attempts to fend off what it refuses to acknowledge, we can already gain a first experience of the nature of the element we are dealing with in the field of power. Here, every instrument, even the instrument of thought, is transformed into a weapon. Here, thought is not concerned with understanding the truth, but with honing its jargon and sharpening its analytical scalpel for the sake of political efficiency. Such instrumentalized thought serves to represent interests that it does not illuminate, and to propagate ideologies that go beyond the tasks of enlightenment, as well as the – undoubtedly justified – self-defence of groups that betray freedom in their approach to their own affects. Methods developed by the critical consciousness to elucidate matters are 'repurposed' and now serve the annihilation of enemies. Under the effect of the 'physics' of the field of power, thoughts are thus twisted into prejudices. One fights by mobilizing affects. Where the affects rule, reason is silent. Philosophical thought can only penetrate the murky flood of passions, interests and ideologies by breaking all the rules that normally apply to human behaviour in politics and society. As a critical thought, it destroys the prejudices on which the self-awareness of the powers and groups in society rely. It opposes agitation with arguments and passion with analysis. It uncovers the hidden mechanisms whereby social power produces false consciousness. In order to achieve this, it must dispense with all weapons, even on the battlefield of power. It must neither disguise itself nor dissimulate. It must reject the solidarity and protection of political groups or ruling oligarchies. It must not ally itself with interests. Its only power lies in its powerlessness.

Philosophy owes the status it has been able to maintain in the domain of European culture, in the face of all dangers, to a magic formula: the inviolable purity of theory. This lived off a carefully guarded belief, or perhaps superstitious belief, in the timeless extra-worldliness of philosophical truth. It was protected by the radiance of the god of metaphysics. Political power operates in time; those who are at home in political reality do not feel threatened in their immediate interests by the recognition of timeless truth. Its magic circle suddenly seems politically vague, and is therefore considered harmless. Metaphysics can even be useful for political powers; it has served, in legitimate and illegitimate forms, to support hierarchies; it has not only elevated the spirit and kept it in motion, but has also provided the costumes for false consciousness. Thus philosophy, despite its repeated confrontations with church orthodoxy, has been tolerated over the last two thousand years because those who held political power in the state and society were generally less perceptive than the fellow citizens of Socrates. Certainly, any philosopher worthy of the name knew that philosophy cannot be socially domesticated. But they were usually wise enough not to make a political virtue of this social necessity.

Lathe biōsas, live in obscurity: this Epicurean motto is a quintessence of worldly wisdom.

Political philosophy is at the centre of Plato's work. In Aristotle, it is transformed – like logic, physics and metaphysics – into an independent philosophical discipline that is constitutive of one of the main areas of philosophy, namely ethics. For the reasons above, however, it is no surprise that politics has disappeared from the classical canon of 'parts' of philosophy since Hellenism. It has moved towards the periphery of philosophical thought. European political theory, despite constant interactions with philosophy, has developed from other approaches. Jurists, historians, publicists, theologians and diplomats rightly did not consider themselves philosophers; on the other hand, aside from a few significant exceptions, philosophy did not conceive any political theories whose theoretical level was comparable to the other philosophical 'disciplines', and which could be placed alongside the political theory of Plato and Aristotle. Only when people wanted to forget, and did forget, what the word 'philosophy' means in the European tradition, could the bad habit of referring to any somewhat extensive political theory thoughtlessly and imprecisely as 'political philosophy' develop. In Hegel, however, politics was pulled into the innermost sphere of metaphysical insight by a new movement of thought. For the first time since Aristotle, something came into being that could be considered a political philosophy in the strictest sense of the term.

Nonetheless, metaphysics indirectly influenced the course of European history to an immeasurable degree. It determined the whole understanding of the world. It already pervaded theology in the time of the Church Fathers. It shaped the forms that were used in the most diverse areas of thought. It was amalgamated in countless refractions, trivialized and popularized, and fashioned in every possible way offered by European culture. Through all vicissitudes and disasters, all political and social upheavals, it remained the bearer of historical continuity and, precisely *because* it was 'metaphysics', that is, the understanding of timeless truth, showed an almost incomprehensible stability in the face of changing historical conditions. The collapse of the Roman Empire, the Migration Period, the crises of the early and late Middle Ages, the Reformation, the age of religious wars and even the 'qualitative leap' of the French Revolution proclaimed by Hegel could not harm it. Only mathematics and the natural sciences are comparable in their stability, and only the development of legal traditions, which was little affected by philosophy, has shown a comparable continuity.

It is to this indirect influence of metaphysics, which can neither be estimated nor measured, that we refer when we accord philosophy, even though it does not directly intervene in the game of power, such great significance also in shaping political conditions that it is justified, despite all reservations, to call the era extending from Plato to Hegel the era of European metaphysics. The two great antipodes of the nineteenth century, Hegel and Nietzsche, agree on this view. There are other, no less legitimate perspectives from which to understand the course of history. If we are speaking of the relationship between philosophy and politics, however, this perspective has a primacy that cannot be overturned. Yet the more decisively one underlines it, the more emphatically one must also note that philosophy as such reflected its interrelationship with politics from Aristotle to Hegel. This is where we find the first manifestation of the 'transcendental semblance' first discovered by Kant, which is produced by the supposed recognition of absolute truth of its own accord.

A clarification of the interrelationship between philosophy and politics forces us to break through this transcendental semblance, and to unmask it even where it has not yet been discovered. In his 'Transcendental Dialectic', Kant limited his critique of transcendental semblance to the topics of classical metaphysics. Today, however, we are in a position to see that the schematic forms of projection he uncovered are just as effective in the realm of experience and the 'positive' sciences as they are in the forms in which metaphysics represents the supersensible. Political thought is likewise dominated by them. One does not change the apparatus of projection as such simply by inserting different templates. The notions to which we give titles like 'capitalism', 'socialism', 'state', 'society' or 'nation' are projected on the screen of public consciousness according to the same rules by which classical metaphysics attempted to establish a supersensible knowledge of God, the world and immortality. Only when we recognize this, only when we are able to break through the thought forms of metaphysics in the realms of empiricism and political praxis too, can we emancipate ourselves from the dominion of metaphysics; only then will we enter the present.

In breaking out of the domain controlled by metaphysics, thought also leaves behind the protective wall with which philosophy, and, under its jurisdiction, all 'pure science', had hitherto succeeded in protecting itself from the blind force of political passions and interests. Now philosophy had to venture out into the field of politics, despite lacking any political weapons and knowing that its own laws will bring it into conflict with all groups. It enters the sphere of social potencies that know how to defend themselves against the power of thought by extremely vicious means. Philosophical thought is always critical thought. This forbids it from showing 'solidarity', as one says today, with political groups of whatever inclination. Whoever philosophizes must fend for themselves. But if one dares to dispense with the protection of political collectives in the realm of politics, one risks more than merely one's personal safety. The further the reach of the ideas one sets in motion, the higher the stakes become.

The crystal enclosure of metaphysics has been shattered. Whoever philosophizes here and now finds themselves cast out into the field of power without allies. Their defencelessness becomes a source of new experience. This transformation begins in the core area of philosophy; we witness it in the 'instrument' of thought, the concept. The crisis of metaphysics unsettles the entire system of preconceptions and prejudices that have shaped today's dominant theories about the nature of the concept. Not only in philosophy, but also in the positive sciences, the same postulate still applies: in academic texts, only concepts that meet the standards of a precise and consistent semantics are adequate. This basic rule of all hard sciences is, as we can recognize today, a mirroring of metaphysics; with the crisis of metaphysics, it loses its justification and its foundation. It proves a secondary derivative of a long-exposed transcendental semblance.[3] If we elude the prejudices that surround this nature reserve of metaphysics with a dense enclosure of class conventions, ritualized rules and magical taboos under the deceptive title 'positivism', those qualities of the 'words', 'terms' or 'slogans' come to light that must be blocked out in order to arrive at an unambiguous semantics. These concealed aspects of the concept reveal something of the 'physics' of the field of power.

In the realm of social and political praxis, concepts are centres of crystallization for diffuse, contradictory and heterogeneous, yet extremely powerful affects. The collective consciousness is set in uncontrollable motion by the use of certain trigger words like 'order', 'violence', 'property' or 'emancipation' in a particular form and in particular constellations with other trigger words. These words derive their power to mobilize consciousness from the historical potencies behind them. Each gained its status in the collective consciousness from the fact that it once served as the watchword for an interest group, a party, a prevailing doctrine or a worldview in a particular political situation. Their mention evokes the energies, now preserved in the collective memory, that were once concentrated in them. The supposed objectivity of science is a result of blocking out the associations that give concepts their 'meaning', of neutralizing them through a sharp-witted system of sanctioned rules, and thus creating the impression that they are unambiguous, although they only arouse interest through the hidden relation to their suppressed ambiguity. This is what we now call 'scientific concept formation'.

Once one sees through the fictitious nature of this approach, the costume of bourgeois decency in which science had managed to dress that most uncanny instrument of humans, the concept, is torn apart. The beast hidden in the concept comes to light. Concepts now manifest themselves as instrumentalized quanta of power. From this perspective, they do not appear in isolation, but rather form manifold constellations with one another according to a previously unknown physics. In each of these constellations they form, through their interference, a medium of consciousness that we (with a metaphor borrowed from physics) have termed a 'field'. The constellations differ according to social groups, generations, interests and so forth. In each of these fields, the same concepts appear in different combinations. They attract, repulse or remain indifferent to one another, and in every field they draw swarms of other 'concepts' into their vicinity like satellites. But tension can also develop between the fields themselves; the consciousness of a society creates itself from the tension between the different fields. What groups or individuals generally discover as the 'content' of their thoughts and reactions results from their respective position within this area of tension.

Philosophy has the task of illuminating the dynamics of the processes from which the collective consciousness emerges. It is exposed to these dynamics, but avoids any identifications that would surrender it to them. It stands in the midst of the game, but does not join in. It struggles to reach its own horizon, step by step, not by

[3] See my introduction to *Offene Systeme II – Logik und Zeit*, ed. Enno Rudolph (Stuttgart: Klett-Cotta, 1981).

misusing critique as a weapon, but by discovering it as a capacity for differentiation, detachment and distance. Without distance, freedom is impossible in both politics and society. Because philosophy arises from freedom and creates freedom, it exists in the midst of society as a constant contradiction of the mechanisms of the collective consciousness. It exists as the antithesis, the objection to the blind game of 'concepts'. It exists by mobilizing the memory of what truly was against the repressions that are constitutive not only of public awareness, but also of so-called science. It exists by destroying, through a critical anticipation of what is to come, the illusions that lead to success in the field of power. It disturbs not only the peace, but also conflicts. Because it does not join in, it is a spoilsport. It does not allow any power to shackle it, for its mission is to remind people that the truth is not subject to humans, but that humans, on the contrary, only have the chance to become human by learning to obey the truth. Exposed in the field of power, it stands in the horizon of truth. Its subject is the difference between power and truth.

The realm of philosophy thus lies in the interstice between the field of power and the horizon of truth. Its task is to break open this interstice and render it visible, for only here can the changing guises of historical reason be explained. The span of this interstice determines the chances of human freedom as such, and of social and political freedom. If this interstice is buried and inaccessible to the collective consciousness, there can be no freedom – regardless of what the states, economic structures and relations of production might look like. Political power always becomes entangled in the net of its own mechanisms. Under the illusion of totality, it changes into powerlessness when it can no longer presuppose freedom and thus deprives itself of the ability to perceive reality as it is through the power of reason. Hence the scope and creative power of political action depend on how much distance it can bear within its sphere. They depend on the intensity and force with which philosophy can bring its powerless objections to bear in the midst of the field of power.

Chapter 8
The Philosophical Concept of Ethics (1978)

Preliminary Note

The international discussion about a 'new economic order' and a more just distribution of our planet's resources has led to ethical demands entering negotiations on economic policy, even economic planning, in hitherto unusual ways. A new relationship is developing between ethics and economics.[1] This has led a study group at FEST to discuss the relationship between ethics and economics from different perspectives. I was given the task of presenting the philosophical concept of ethics. This was significant for the topic because the connection to economics was more important for the development of philosophical ethics in the European tradition than is generally known.

The Babylonian confusion of our time is also reflected in the range of often colourful meanings that are attached to the word 'ethics'. If one tries to gain an overview of the dominant views of ethics today (see, for example, the entry 'Ethics' in the *Historical Dictionary of Philosophy*), one is left with the impression that completely disparate things are being discussed under the same title, but that barely anyone still asks to what the name actually refers. One finds almost equally divergent views about what economics is and what – and ethics already makes an appearance here – it should perhaps be. In this intellectual situation, any attempt at a convincing synthesis of ethics and economics opens itself up to the suspicion of deception or self-deception. At any rate, we must be aware that we live in a global economy, but that both our economic theory and our ethical reflections represent specifically European traditions; this applies equally to the United States and the Soviet Union, albeit in a different sense. Our thinking, even when it seems to distance itself from metaphysics, is formed by European metaphysics and Christian theology, as well as

---

[1] This text was published first as: "Zum philosophischen Begriff der Ethik". In: Ch. Frey/W. Huber (eds.), *Schöpferische Nachfolge: Festschrift für H.E. Tödt*. Heidelberg: Texte und Materialien der Forschungsstätte der Evangelischen Studiengemeinschaft, Reihe A Nr. 5 (1978), pp. 61–94. Also in: *Hier und Jetzt* Vol. I, pp. 137–164. The permission to republish this text here was granted by Klett-Cotta.

E. Rudolph and J. Picht (eds.), *Georg Picht: A Pioneer in Philosophy, Politics and the Arts*, Pioneers in Arts, Humanities, Science, Engineering, Practice 19,
https://doi.org/10.1007/978-3-030-31790-4_8

the crisis of these two sciences[2] in the nineteenth and twentieth century. As we have neither adopted nor processed the ethical wisdom of other cultures and religions so far, the legitimacy of the ethical assessment of the global economic problems of our time rests on a fragile foundation. The least that can be demanded of us in the context of the incipient global conversation is that we once more find a way to account for our own presuppositions. The following sketch attempts to outline in brief what the science that chose this name means by 'ethics'. I have specifically restricted myself to the field of philosophy.

I

> Ancient Greek philosophy was divided into three sciences: *physics*, *ethics* and *logic*. This division is perfectly suitable to the nature of the subject and there is no need to improve upon it except, perhaps, to add its principle, partly so as to ensure its completeness and partly so as to be able to determine correctly the necessary subdivisions.
>
> All rational cognition is either *material* and concerned with some object, or *formal* and occupied only with the form of the understanding and of reason itself and with the universal rules of thinking in general, without distinction of objects. Formal philosophy is called *logic*, whereas material philosophy, which has to do with determinate objects and the laws to which they are subject, is in turn divided into two. For these laws are either laws of nature or laws of freedom. The science of the first is called *physics*, that of the other is *ethics*; the former is also called the doctrine of nature, the latter the doctrine of morals.[3]

Kant placed these paragraphs at the start of his *Groundwork of the Metaphysics of Morals* in order to assign to this text the place it occupied within his system of a metaphysics of finite reason. According to the conception of this system from the chapter on the architectonic in the *Critique of Pure Reason*,

> Now the legislation of human reason (philosophy) has two objects, nature and freedom, and thus contains the natural law as well as the moral law, initially in two separate systems but ultimately in a single philosophical system. The philosophy of nature pertains to everything that is; that of morals only to that which *should be*.[4]

Metaphysics is therefore divided 'into the metaphysics of the *speculative* and the *practical* use of pure reason, and is therefore either *metaphysics of nature* or *metaphysics of morals*.'[5] These two disciplines of metaphysics correspond to Greek physics and ethics. In Kant, logic is transformed into transcendental logic, which replaces ontology. It holds within itself the groundwork for the whole system, and

---

[2] 'Science' should not be understood here in the widespread and characteristically modern sense, restricted first to the natural sciences, based on empirical data and communicated in the language of mathematical logic, and then to the slightly wider field of sciences in the similarly empirical areas of society and politics. Rather, science (*epistēmē*) is here the form in which valid and retrievable knowledge of a specific, already well-defined subject area is gathered. Such areas also include the field of what is conceivable but not concretely given, such as theology (*epistēmē theologikē*) or the field of what is not, but should be (*epistēmē ēthikē*) (ed.).

[3] Immanuel Kant, *Groundwork of the Metaphysics of Morals*, ed. and trans. Mary Gregor (Cambridge University Press, 1997), p. 1.

[4] Immanuel Kant, *Critique of Pure Reason*, ed. and trans. Paul Guyer and Allen W. Wood (Cambridge University Press, 1998), p. 695.

[5] Ibid., p. 696.

simultaneously fulfils Kant's requirement that the principle of dividing philosophy into its three disciplines should already be contained in the groundwork for the system. Because transcendental logic shows the condition of possibility for both a metaphysics of nature and a metaphysics of morals, Kant's conception characterizes it as a 'metaphysics of metaphysics' (letter of 11 May 1781 to Markus Herz). It is the science of the self-knowledge of pure reason; thus it is identical with its critique (*Critique of Pure Reason*, A XI/XII).

The philosophical science that Kant, referring to the ancient Greeks, calls 'ethics', is – as the conception of his system shows – something completely different from the specialized discipline that established itself in university philosophy during the second half of the nineteenth century under the same name. The three sciences of physics, ethics and logic are meant to encompass the totality of what can be thought, such that this division assures us of their completeness. The division is meant to occur according to a principle, that is, it is meant to be given a priori for reason, universally valid and necessary. It is only through our understanding of this principle, according to Kant, that reason 'comes to reason' in the first place. The principle of this division constitutes that identity of reason with itself on which both its truth and its freedom are based. It is not, Kant argues, for us to choose what we consider 'ethics' and how we wish to demarcate its field. The 'cosmos of beings' predetermines the boundaries in which the possibility of reason as such is constituted by the possibility of the knowledge we call 'ethics'. That is why Kant emphasizes time and again in both the *Groundwork* and *Critique of Pure Reason* that ethics is concerned not only with the human being, but with the principles of reason as such. Kant was convinced that there are living beings capable of reason, with an entirely different constitution, on other heavenly bodies in the universe. They would have other rules of practical behaviour corresponding to this constitution, but he considered it an apodictic certainty that the categorical imperative would apply to them too. As metaphysics, the metaphysics of morals asserts a universal claim. The categorical imperative is a cosmic principle; the transcendental subject of our reason is identical to the transcendental subject of reason for all living beings in the universe that are capable of reason. It is, one might say, the transcendental world soul. Kant's 'ethics' is the science of the principles through which this universal reason defines itself. That is why 'the universal imperative of duty can also go as follows: act as if the maxim of your action were to become by your will a *universal law of nature*.'[6] A universal law of nature is a law that applies to the entire universe.

According to Kant's conception, the forms in which the principle of freedom asserts itself under the conditions of temporal and sensual human existence only belong in the empirical part of ethics, namely 'practical anthropology'.[7] This encompasses all forms of human behaviour that are inconceivable without freedom: politics, economy, law, morals and all expressions of individual life. The totality of so-called 'human sciences' are sub-disciplines of this 'empirical part' of 'ethics'. Kant's philosophy of law and his philosophy of history are the two main areas of

[6] Kant, *Groundwork of the Metaphysics of Morals*, p. 31.

[7] Ibid., p. 2.

'practical anthropology'. The counterpart to the discipline we now call 'ethics' is found neither in the *Groundwork of the Metaphysics of Morals* nor the *Critique of Practical Reason* – rather, as 'groundwork' for the *metaphysics* of morals, they are parts of transcendental logic. What Kant would have taught if he had been expected to give a seminar on 'ethics' in the sense that is customary today can be found in his lecture on anthropology. In Kant's system, the philosophy of law has the function of mediating between nature and freedom. The principles of such a mediation likewise have universal significance; that is why the metaphysical theory of law forms the first part of the *Metaphysics of Morals*. The philosophy of law is meant to show a priori how the *existence* of freedom, its manifestation in time, is possible. It is therefore of crucial importance for the entire metaphysics of morals. That this is too seldom acknowledged can be explained by the circumstance that Kant, in order to simplify the presentation of his ideas, structured the *Groundwork* and the *Critique of Practical Reason* in such a way that they only refer to the second part of the metaphysics of morals, namely the metaphysical theory of virtue.[8]

In the preface to the second edition of the *Critique of Pure Reason*, Kant says that among 'that admirable people the Greeks',[9] logic and mathematics already went through the revolution of thought that would show the secure path of a science for all time; natural science, however, only discovered the 'highway of science' with Bacon, Galileo and Torricelli.[10] Things were even worse for metaphysics, whose methods prior to the *Critique of Pure Reason* were 'a mere groping, and what is the worst, a groping among mere concepts.'[11] This makes it all the more amazing how emphatically he declares that the ancient Greek division of philosophy (i.e. metaphysics) into physics, ethics and logic is 'perfectly suitable to the nature of the subject', and that 'there is no need to improve upon it except, perhaps, to add its principle'. For with the division of philosophy, as he knew full well, Kant adopted from the Greeks the outline of all that can be thought which dictated the dimensions of European thought from Aristotle to Hegel. Kant does not want to throw the Greek science of metaphysics over board and replace it with a new science of the same name. His 'Copernican revolution' is merely to consist in 'adding' the principle of this science. He wants to make the terrain for the 'majestic moral edifice' of Platonic philosophy 'level and firm enough to be built upon, for under this ground there are all sorts of passageways, such as moles might have dug, left over from reason's vain but confident treasure hunting, that make every building insecure.'[12] Thus Kant himself points out that we must turn back to Greek philosophy if we are to understand the original purpose of the division that gave Europe the concept of 'ethics'.

---

[8] On the relationship between the philosophy of law and the philosophy of history, see 'Kant's Transcendental Grounding of International Right', pp. 75–105..

[9] Kant, *Critique of Pure Reason*, p. 107.

[10] Ibid., p. 108.

[11] Ibid., p. 110.

[12] Ibid., p. 398.

II

The division to which Kant refers came about in Plato's Academy, where its originator was Xenocrates, the friend and ally of Aristotle. One can only understand it if one appreciates what was actually being divided. We find the answer in the word that requires the additions of *physikē*, *ethikē* and *logikē*: *epistēmē*, 'science'. In this division, philosophy views itself as the science of human knowledge as such; it is the science of science. Each individual science proves its possibility of being a science, that is, its truth, by the fact that it can be attributed to this division. That philosophy can view itself as science, in an explicit contradiction of Plato, is due to the fact that Xenocrates, like his friend Aristotle, taught that God, the origin of all knowledge, is pure *nous*; he is not located beyond being, like the Platonic 'idea of good', but is himself, as pure substance, the *summum ens*. The human being can have knowledge because the soul has the capacity of the *nous pathetikos*, which becomes perfectly identical with the divine *nous* in the act of recognizing the truth. If that is the case, then we can no longer accept Plato's characterization of philo-sophy, the *striving* for knowledge, as the highest form of human cognition; for then it is possible that we can achieve knowledge itself. Hegel adopted this joint programme of Xenocrates and Aristotle in the preface to the *Phenomenology of Spirit* with peculiar emphasis. The division of philosophy into the three basic disciplines of physics, ethics and logic necessarily arises from the decision that philosophy, which only becomes 'metaphysics' in this way, must have the form of science. In Plato's philosophy, it would be impossible to divide the infinite movement of thought that he called 'philosophy' into disciplines. The transition from philosophy to science, however, put thought in a position where the whole of what can be thought could only be presented in its full unity by dividing it. For Xenocrates and Aristotle too, the 'object' of philosophy, as for the pre-Socratics and Plato, remains *tode to pan* – all of this here, the universe. But the unity that makes the universe this universe can now only be revealed by distinguishing three dimensions in which all human thought, whatever its object, necessarily operates. The titles of 'physics', 'ethics' and 'logic' thus point towards the three dimensions in which science recognizes the unity of the universe. This does not involve three different 'areas' of possible knowledge being separate from one another. Rather, the purpose of the distinction is to show that each individual science stands within the horizon of physics, the horizon of ethics and the horizon of logic, and only attains the rank of a science when these three dimensions of all possible knowledge can be brought into alignment. For only then can all three titles be connected to the singular *epistēmē*.

I will attempt to explain this in greater detail:

1. Everything that can be known at all, including humans, gods and numbers, is still considered a phenomenon within *physis*. If one only considers what is known in a particular science, its so-called 'subject area', then everything knowable belongs in the realm of physics.
2. But every science is at once an art of the human being. Therefore, as Aristotle shows in Chapter I of the *Nicomachean Ethics*, it is a link in a hierarchy of

purposes that culminate in recognizing, for humans, the nature of good. If one views science from the perspective that it is humans who are to possess or acquire knowledge, then every science, including 'abstract' sciences such as mathematics, is a part of ethics.

3. Humans can recognize the truth, but they can also be wrong. All human knowledge is deceptive; to turn opinions into knowledge, one must discover the right way to understanding. In Plato's school, *logos* was the technical term for a 'straightforward examination'. The science of the correct ways of understanding, which Plato first termed *methodos*, is called 'logic'. As no science is possible without a 'method', every science, whatever it examines, belongs to logic.

The fact that every possible science belongs to the domains of physics, ethics and logic means that the triad of physics, ethics and logic is what constitutes every individual science as such in the first place. As soon as we recognize and know this, it results in a second-order science, the science of science on whose possibility the young Plato already reflected in the *Charmides*. It must be a recognition of the three-dimensionality of all knowledge. As each of the three dimensions demands its own direction of thought, its own 'method' and thus its own discipline, the science of science can only be built up by being divided into the three basic disciplines of knowledge. It is only the distinction between these three mutually irreducible basic disciplines and the differentiation between their dimensions that give thought the clarity which makes it capable of science.

This initially very simple outline of philosophy as science holds problems that Aristotle already brought to light and thought through with penetrating brilliance. Working from a different perspectives, the Stoics would later make repeated attempts to discuss the meaning and internal problems of this distinction. There is no need to concern ourselves with that here; but we do need to ask why these three disciplines in particular should contain the totality of possible human knowledge. This is explained by the pre-Platonic history of Greek thought.

III

The writings and didactic poems of the great pre-Socratic thinkers, Heraclitus and Parmenides, Empedocles and Anaxagoras, were – we do not know exactly when – given the title *peri physeos*. There was only a single philosophy here: the theory of *physis*, namely 'physics'. In order to understand this title, one must realize that in Greek thought, humans and gods were also part of *physis*. That is why theology appears in Aristotle's *Physics* as the final sub-discipline of *physikē epistēmē*. If *physis* can be thought and understood in this way, there is no need for any division in order to have an overview of the totality of philosophy; then philosophy is no different from physics, and there is no space for the separation of an independent 'ethics' or 'logic'. But this first conception of philosophy was derailed by Sophism.

Protagoras made the famous statement that 'Man is the measure of all things: of the things that are, that they are, of the things that are not, that they are not.' But this is explained by a similarly constructed statement from another text: 'Concerning the gods I cannot know either that they exist or that they do not exist, or what form they might have, for there is much to prevent one's knowing: the obscurity of the subject and the shortness of man's life.' The second statement justifies the cool-headed scepticism of the first. Mortality sets the boundary that prevents humans from recognizing the gods and everything else that is in *physis*. Mortality gives humans their measure, and this measure does not correspond to the truth. Hence we only grasp a thing as it appears to us in keeping with our measure: 'Things *are* to me as they *appear* to me' (Plato, *Theaetetus*, 152a, in his explanation of the homo-mensura principle). According to Protagoras, this measure has its own perspective: we grasp everything that shows itself to us in terms of how we can use it. Our view of reality is crucially determined by our need (*chreia*). Hence 'things' do not appear to us as the existent that they are in their own right, but as *chrēmata*, objects of use. This distinction between *on* and *chrēma* is the first version of the later distinction between the thing-in-itself and the appearance. As the perspective from which we judge the *chrēmata* stems from what we consider good, the science of what is good for each human determines our forms of perception and our perception of nature. This science rests on a knowledge of the manifold *ēthē*, the morals, customs and behaviours of different peoples. Depending on region, climate, traditions and other conditions of life, every people and every tribe has its own *ēthē*. For some, 'good', namely that which serves their preservation, is one thing, and for the others it is another. Anyone who wishes to understand the world and know humans must investigate these *ēthē*. The science of the relativity of different 'value systems', that is, the 'behavioural research' of a 'cultural anthropology' premised on the homo-mensura principle, should bear the name 'ethics'.

Protagoras gave the text containing the homo-mensura principle the title *Truth* or *Refutations*. What he wanted to cast down was the 'physics' of the great pre-Socratics. By contrast, his sceptical resignation, which does not presume to break out from the confines of mortality, is, as could easily be shown, in full agreement with the fundamental convictions of popular religion. He was not the 'atheist' as which he was later presented. His statement speaks of the boundaries, not the unconditional potency of humans. It was only in Modern Age Europe that the statement became a slogan that twists its meaning into the very opposite: now humans proclaim themselves the subject of all that can be known, and use this self-definition to claim total control over everything in nature that is accessible to them as *chrēma*, as raw material for their production.

The scepticism of Protagoras marked a shift in thinking that could not be reversed. When Plato and Aristotle sought once more to find the truth, they would have to deal with this doctrine of the finitude of human knowledge. Thought could no longer define itself as the immediate understanding of *physis* as a whole; it had to reflect on the fact that everything which presents itself as knowledge is knowledge known by humans. Thus the understanding of humans and of what good means for them became an intellectual task of its own alongside the old physics. For the same reason, however, knowledge as such and its possibility became a philosophical issue with its own structure that could not be accommodated either by the framework of physics or the framework of the science of the human being. This explains the emergence of the three basic dimensions of all thought: physics, ethics, logic.

The doctrine articulated by Protagoras rests on an anthropology that we know thanks to Plato. In the dialogue named after Protagoras, Plato has him recount a myth whose form is that of a Platonic work of art and whose content is a faithful reproduction of the teachings of this Sophist, whom Plato regarded with great respect (320c). According to this anthropology, Epimetheus's lack of foresight made humans 'deficient beings'.[13] They would not have been capable of living if Prometheus had not stolen fire from the gods, along with the associated skills, and equipped them with this. Now, however, humans are bringing about their own downfall; they destroy one another because they are unable to live as a community. In order to save them from demise, Zeus sends them, via Hermes, *aidōs* and *dikē* – reverence and justice. Now they learn to live in a polis. Their preservation is ensured.

The following aspects of this myth should be noted to elucidate the emergence of the concept of ethics:

1. *Chreia*, need, is fundamental to human understanding of the world because the human beings is by nature a deficient being.
2. The demiurgic nature of *technē*, which already contains important aspects of later 'technology', cannot ensure the preservation of humans because it can just as easily be used for the purpose of annihilation. It is indispensable, yet at once blind to what benefits humanity.
3. Human beings owe their preservation to ordered coexistence in a polis. But this, according to Protagoras, is not made possible by any knowledge, by the cognition of divine norms, for such a knowledge would go beyond the measure imposed on humans by their mortality. Rather, Zeus gives humans the basic attitudes of *aidōs* and *dikē*, which can neither be invented nor artificially produced by them, and are not under their control; they are honoured as divine powers. This means that the later 'ethics' cannot be separated from 'politics', the art of living in a polis, and that its central theme is that which evades the control of humans and limits their actions to the boundaries set for them. Ethics is not a 'theory of action', but rather the theory of the intransgressible limits of action. In terms of its historical origins, it is a doctrine of the finitude of human knowledge, and simultaneously has the task of showing why humans pave the way for their own downfall if they are unable to see or unwilling to admit to the limits of their knowledge. Despite his opposition to Protagoras, Plato always upheld this belief.

IV

This is not the place to expound the ethical content and architecture of the teachings of Plato and Aristotle. I will restrict myself to examining the location of economics in both conceptions.

[13] Gehlen famously based his anthropology on this principle, transmitted via Herder and Nietzsche and biologically underpinned by Portmann.

Following on from Protagoras, Plato attributes the genesis of the polis to the need of humans and the resulting necessity of a division of labour – that is, to the terms of any possible form of human existence, which require them to have an economy and force them to work. Like Protagoras, he refers to the ancient wisdom of Hesiod, who, like the Bible, taught in a famous and much-quoted statement that 'the gods have placed sweat before excellence [*aretē*]'. It is the struggle against need, namely work (in the broadest sense of the word), that shapes the habits and customs, the ways of life and behaviours, the *ēthē* of humans, depending on the place they are assigned. These are different among hunters or shepherds, among farmers or craftsmen or warrior; each of these groups, corresponding to its specific work, has its own specific *aretē* to which all other behaviours are assigned. *Aretē* did not mean for the Greeks what later moral philosophy called 'virtue'. The *aretē* of the cow is that it gives milk, that of the horse is its speed, that of the dog is its watchfulness and that of the knife is its sharpness. *Aretē* therefore means that fitness to fulfil a purpose which is 'naturally' assigned to a thing or creature because it makes it what it is. As every thing and every creature exists only by dint of its *aretē*, this alone guarantees its preservation. The striving to develop that specific *aretē* is therefore identical to the striving for preservation. To the extent that later ethics was a 'theory of virtue', it also always had to be a theory of the conditions of human preservation. The first prerequisite for the preservation of humans, however, is their economy, which is only possible in the framework of a political order. What is meant by 'economy' is conveyed by the word itself: *nemein* means 'assign', *oikos* is the domestic community; *oikonomia* is the providing household management that assigns to each person their individual work and what they are entitled to for the satisfaction of their needs within the framework of preserving the domestic community as a whole. ('Economics', then, is not the science of business, but organized business itself.) This structure was transferred to the polis. The entire conception of the Platonic polis is erected on this 'economic' ground plan, and Aristotle followed Plato in this. It still acted as a model for Hegel's philosophy of law, and explains the status that the 'system of needs' has in his deduction of the state.

So economics appears in ethics in two ways. On the one hand, it determines the conditions under which the many varieties of human *ēthē* develop; there are no forms or rules of behaviour that are unaffected by economic conditions. Thus economics is already present in the subject matter of the 'science of behaviours' (ethics). On the other hand, we need such a science because humans are constantly faced anew with the question of which forms of economy are beneficial for their preservation and for the development of their *aretē*, and which forms rather have a destructive effect under the mask of utility. In both directions, ethics entails an understanding of the natural conditions under which the deficient human being has to assert its existence. The primary subject area of ethics is therefore not human consciousness; it grows from the constitutive determinants of human existence in nature. On the other hand, economics is not a sector of social practice that is indifferent to ethics to which one could 'apply' an ethics born of entirely different motives; rather, in addition to material goods, the *ēthē* too are economically produced and feed back into economic production. Ethics is an integral part of an economics that sees through its own mechanisms. The science of *ēthē*, ethics, presupposes a particular ethos that, as Aristotle shows in Chapter I of the *Metaphysics* (981b13ff.), can only develop under specific economic conditions. The knowledge claimed by ethics can only be transparent to itself if it comprehends how everything it teaches refers back directly or indirectly to economics.

All these definitions are based, as we have seen, on the fact that need (*chreia*) governs the lives of humans. In the soul, the counterpart of need is the faculty of desire (*epithymia*). This faculty is the fundamental determinant of life as such, as living species preserve themselves through nourishment and mating. The power of the drives that originate directly or indirectly from the faculty of desire is explained by the omnipotence of *chreia*. But humans, thanks to their demiurgic arts, can escape naked need and set up complicated cultural systems that open up completely new fields and possibilities for their desire. Plato begins with this idea in the *Republic*. He shows that the faculty of desire knows neither measures nor boundaries of its own accord, but that, wherever it can develop its own dynamics, it pushes towards a boundless expansion of needs and the instruments of power necessary to realize them. This leads outwardly to war, and inwardly to a battle of interests that inevitably brings about the collapse of the state. As long as the faculty of desire is directed purely at the satisfaction of natural needs, it is limited by saturation. Emancipation from these limits is made possible by the invention of money; money can be accumulated ad libitum. The accumulation of money is accumulation of power. In addition, money allows an unlimited interchangeability between the aims of the faculty of desire. For Plato, *ploutos*, wealth, stands for everything that the faculty of desire considers useful and beneficial, and therefore good. Yet it is precisely the striving for *ploutos* that also causes wars and the self-destruction of the state: that which *appears* good is in reality pernicious.

Here we must introduce a new basic concept from Greek ethics: that of *bios*. In the context of ethics, this word refers not to organic life, nor the life of an individual, but rather the type of a particular form of life whose totality is shaped by what it deems worth striving for. *Bios* is therefore the overarching term for a rich and manifold complex of *ēthē*. The *bios* that gains its character from striving for wealth is called *bios chrēmatistikos*. One of the basic teachings of Plato's *Republic* is that any polis will collapse as soon as the representatives of *bios chrēmatistikos* gain such power that money dominates all forms of life that are still possible in such a polis. In Plato's analysis, this caused the fall of Athens. The subject matter of the science of behaviours then consists of the desires that emancipated themselves from immediate needs in the wake of higher cultural development (today we call their non-sublimated form 'interests') and power. The art of dealing with power is called 'politics'. As long as ethics, economics and politics are practised as separate 'disciplines', that which should be known in these sciences necessarily remains concealed.

Based on this analysis, one can no longer say that the human being is the measure of all *chrēmata*. To the extent that they are driven by their faculty of desire, humans are rather beyond measure, beyond moderation, and that is what causes their downfall. They cannot set the measure themselves, but must recognize it as set for them. Limits are set for humans within which they are able to develop themselves through their position in *physis* towards *physis*. They must therefore recognize *physis* as a whole if they wish to recognize what is good, beneficial and wholesome for them. Thus ethics is forced by its own logos, that is, through the straight path of examining what humans should consider good, to undergo a transition to physics. But it can only perform this transition by finding a method that allows the discovery of the truth of *physis* for humans, despite the boundaries of their mortal knowledge – no longer directly, but by recognizing these boundaries. This reflection shows that the three basic disciplines of philosophy – ethics, physics, logic – are held together by the shared question as to what good actually is, that is, the 'idea of good'. In the *Laws*, Plato calls the idea of good *theos*. This explains the statement: 'So God should be the measure of all *chrēmata* for us to the highest degree, far more so than any human, as is claimed by some' (716c). – What does 'measure' mean now?

From the perspective of the faculty of desire's urge to expand, measure simply appears as a 'limit' to growth. An expansion with no self-regulatory mechanism to restrain it from within can only be limited by coming up against *external* limits. For Plato, however, the search for the truth of *physis* leads to the discovery of *internal* proportions that keep everything that exists in a state of being. The word 'measure' then refers to the unstable equilibrium of an immanent structure. The preservation of humans depends on discovering and maintaining this movable equilibrium of the correct measure, both in their soul and in the state. Only from the context of physics can we explain why Aristotle, following the model of Plato, developed an ethics that is quite simply a theory of proportions in a complex system. Although Aristotle treated the science of *ēthē* and the science he termed 'politics' (which includes economics) in separate books, he leaves no doubt that these two sciences form a unity. He was led by purely methodological factors to lay out his ideas in such a way that the ethics and his investigations of politics would be literarily independent of each other. One sees here, as elsewhere, that the methodological concepts of his theory of science – his 'logic', in the broader philosophical sense of the word – sometimes collide with the contexts that arise from the issues themselves in ways that have far-reaching consequences. An isolated, specialist discipline called 'ethics' would probably never have come into existence if Aristotle had followed different methodological principles in the disposition of his books. Once the – only ostensibly – isolated science of 'ethics' existed, the laws of the scientific division of labour made it inevitable that it would henceforth be pursued as a specialist discipline; but that was no longer philosophy, and is therefore not part of our subject here. Let us note what Kant confirmed: that, in the triple division of 'physics, ethics, logic', the concept of 'ethics' has the broad meaning whose emergence and purpose have been outlined here.

V

If we make a leap of over two thousand years, the time separating Kant from Plato and Aristotle, what is initially surprising is the continuity. Kant and Hegel, in Königsberg and Berlin, found themselves confronted with the same problems between 1780 and 1830 as Plato and Aristotle in Athens between 380 and 330 BCE. They attempted to resolve them within the same basic intellectual order encapsulated by the triad of physics, ethics and logic. In the meantime, empires have been shattered, religions founded, new worlds discovered; mass migrations, revolutions and wars ploughed up societies; the world spirit emigrated from the Aegean to the Baltic Sea and had to adapt to incomparable economic conditions. All the historical substrates we tend to view as 'substructures' of intellectual constructs were exchanged. All this was not enough to shake the edifice of European metaphysics, however. This can only be explained if this edifice, beyond changes in historical conditions, is based on the phenomena themselves, or, as Kant says, on 'the nature of the matter'. Anyone seeking to understand 'ethics' must realize that there is no more room for mere arbitrariness here than in physics. The fluctuations of opinions, ideologies and theoretical fashions do not touch the bedrock of basic conditions that ethics must deal with. We will therefore begin by following this continuity so that we can determine later where Kant's moral philosophy differs fundamentally from the Greek conception he adopted.

Like Plato, Kant teaches that humans have a natural disposition towards ethics because human action is guided by the faculty of desire. He too considers the faculty of desire a fundamental element of life as such: 'Life is the faculty of a being to act in accordance with the laws of the faculty of desire.'[14] In contrast to theoretical knowledge, which focuses on what is, we desire that which is not, but which we want to be. Hence the faculty of desire is the faculty of the living being 'to be by means of its representations the cause of the reality of the objects of these representations'.[15] This definition is not restricted to humans, but applies to living beings as such. The difference between human and animal lies in the fact that for an animal, 'the whole rule of its conduct would be marked out for it far more accurately by instinct'.[16] It therefore does not need to be aware of its representations as such. The human faculty of desire, on the other hand, is subject to the 'weak and deceptive guidance' of reason. Consequently, humans are aware of these representations and find themselves in the worrying situation of 'meddling with nature's purpose'.[17] They pay for freedom by being at the mercy of their own fallibility: 'Now in a being that has reason and a will, if the proper end of nature were its *preservation*, its *welfare*, in a word its *happiness*, then nature would have hit upon a very bad arrangement in selecting the reason of the creature to carry out this purpose.'[18] Ethics is necessary because the rationality of humans themselves must first be brought to reason; otherwise it will ruin them. Here too, Kant concurs with Plato.

By virtue of their rationality, humans have the freedom to direct their faculty of desire at individual purposes as they choose and to subordinate their other drives to the priority of these purposes; they can determine their faculty of desire. In so far as reason determines the faculty of desire (and thus life), it acts as practical reason or as will. It is free. To the extent that it obeys needs, however, it is subject to the dependencies

---

[14] Immanuel Kant, *Critique of Practical Reason*, ed. and trans. Mary Gregor (Cambridge University Press, 1997), p. 7n.

[15] Ibid.

[16] Kant, *Groundwork of the Metaphysics of Morals*, p. 9.

[17] Ibid.

[18] Ibid., p. 8.

of our sensual existence in time. It is unfree. All conscious human action in time (in contrast to instinctive behaviour) must bridge the irresolvable antinomy between its freedom and its dependency. It must therefore impose boundaries on how the faculty of desire sets its goals. Since the faculty of desire acknowledges no boundaries, humans require a legal system for their preservation that, through its sanctions, forces them to act as their true interest, in contradiction of the wilfulness of their desire, must demand. Right is a substitute for instinct. The schematicisms on which it draws, in analogy to the mechanism of instincts, are known as 'norms'. Norms do not appear in Platonic-Aristotelian ethics, which deals not with laws but with proportions; this corresponds to the Greek doctrine of *physis*. The physics of the Modern Age thought functionally: it knew units of measurement, but not measures. It replaced the equilibrium of proportions with the law. That is why the first part of ethics, the metaphysical doctrine of right, which is supposed to mediate between freedom and nature, had to become a theory of laws – that is, a normative science.

All action in which the faculty of desire is guided by rationality is oriented towards an end that requires the provision of means. It follows a formula: if you want a and b, you must do x and y. Kant calls this formula the 'hypothetical imperative'. (Here the word 'hypothetical' has the same meaning as in 'hypothetical' clauses in grammar.) Whether I want or should want what is named in the conditional clause remains open. But if I do want it, I must act as stated in the consequent. This asymmetry between wilfulness and necessity defines the entire existence of humans in time. The basic formula of hypothetical imperatives thus regulates all human action, and economics in particular. Every form of purposeful action, of planning, is subject to this formula. It has the same fundamental significance as the analogously constructed fundamental principle of synthetic judgements in the realm of theoretical cognition. The problem of ethics can now be specified: it is meant to convey a knowledge that enables us to formulate the conditional clauses of hypothetical imperatives such that they agree with the central purposes of human nature. The nature of the human being lies in its reason. If hypothetical imperatives are to align with reason, reason must first be aligned with itself. Kant refers to the formula that states the conditions under which reason can be in agreement with itself as the 'categorical imperative'. Through it, reason defines itself out of the freedom to be identical with itself. In this sense it is 'auto-nomous'. The concept of autonomy in Kant has no 'moral' meaning, but rather a metaphysical one. It states that freedom is the condition of possibility for the identity of reason with itself, and thus for the cognition of truth.

The science of the conditions 'under which the understanding can and ought to agree with itself alone'[19] is logic. Hence, according to Kant, the categorical imperative, like the formula of hypothetical imperatives, belongs in the area of logic – indeed, it is even the condition of possibility for logic. Here, reason is no longer concerned with the matter of the faculty of desire, but only with itself: 'All rational cognition is either material and concerned with some object, or formal and occupied only with the form of the understanding and of reason itself and with the universal rules of

[19] Immanuel Kant, *Lectures on Logic*, ed. and trans. Michael J. Young (Cambridge University Press, 1992), p. 529.

thinking in general'.[20] As both the metaphysics of nature and the metaphysics of morals can only be grounded in the reason that conceives of these sciences, ethics too must be derived from the pure self-knowledge of this reason, that is, from transcendental logic. Transcendental logic is therefore the science of the subject of ethics, the subject as such. The categorical imperative states that the 'self' in which reason recognizes itself cannot be thought as something that 'exists' in whatever form, but can only be by defining itself in the alignment between our sources of knowledge for the infinite task of its identity. This forms the centre of Kant's theory of freedom.

If one compares Kant to the Greeks, it transpires that the freedom envisaged in the logical form of the categorical imperative replaces Plato's assertion that God is the measure of all *chremata*. In the unconditional awareness of freedom, reason posits its own imperative as a measure. Protagoras triumphed over Plato. That, at least, seemed the obvious conclusion after the concept of the subject had been restricted to humans in the nineteenth and twentieth century. For Kant, however, subjectivity is a universal idea. It 'begins from my invisible self, my personality, and presents me in a world which has true infinity but which can be discovered only by the understanding, and I cognize that my connection with that world [...] is not merely contingent [...], but universal and necessary.'[21] I am connected to it through those 'worlds upon worlds and systems of systems'[22] that fill the visible universe. According to Kant, then, the idea of freedom takes humans out of themselves. It provides the measure for mere being-human in time through true, intelligible infinity.

Because reason, in so far as it is in agreement with itself, has the authority to define the conditional clauses of hypothetical imperatives as if they were unconditional clauses, it is sovereign in the field of practical anthropology. It knows only one standard: its agreement with itself. Once the subject has constituted itself in this way, this has unforeseeable effects on the forms in which its will determines the faculty of desire. In both economic and political actions, human beings demand the unconditional sovereignty of a reason that sees itself as an absolute subject, even though, because of the purposes of their actions, they are at the mercy of the heteronomy of the faculty of desire. The *Homo economicus* of the nineteenth and twentieth century treats hypothetical imperatives as if they were categorical imperatives; the subject, even where it is dependent, demands unconditional freedom. It projects its absolute self-awareness onto the field of heteronomy and thus produces the paradoxical phenomenon of an absolute heteronomy.

The problem behind the ambivalence of the Kantian imperatives can be very simply formulated: 'The philosophy of nature pertains to everything that *is*; that of morals only to that which *should be*.'[23] We call the agreement between our cognition and the existent 'truth'. That which should be is defined by the fact that it is *not*. Utterances that obey the 'logical' form of the imperative – wish, request, demand, command – were excluded from the realm of logic by Aristotle because, following

---

[20] Kant, *Groundwork of the Metaphysics of Morals*, p. 1.

[21] Kant, *Critique of Practical Reason*, p. 133.

[22] Ibid.

[23] Kant, *Critique of Pure Reason*, p. 695.

the law of non-contradiction, it is not possible to claim that they are true or false. The proposition 'All bodies are extended' is true. The proposition 'All humans are equal' does *not* correspond to the given state of affairs; it contradicts reality and draws its moral power precisely from this contradiction. If ethics refers to that which only *should* be, it refers automatically to that which is *not*. The truth it claims denies true theoretical cognition. It is part of its nature that it does *not* correspond to reality. Like Plato, Kant concluded from this that the objective reality to which cognition relates in true or false judgements is only an appearance; for if it were the truth in itself, then freedom would be untrue and hence impossible. He was consistent enough to teach at the same time that the form of understanding that is the judgement cannot be used in the field of freedom. Whenever we use the logical form of the judgement in ethics, we fall prey to transcendental semblance. That is why the 'fundamental law of pure practical reason' does not have the logical form of a principle; it is rather a commandment (*Critique of Practical Reason*, § 7). A principle is a general major premise that applies to all individual cases. The unconditional nature of the commandment rests on a different form of synthesis; here, universality has a different form. 'Do not kill' is not a universal norm. The proposition is not equivalent to the proposition 'No one kills'; rather, it aims at the individual and demands of them that they, as an individual, subordinate their actions to a maxim that would have to be the principle of a universal law, a norm, if freedom were reality – and would therefore not fall into the area of what simply should be. We find the same structure in all tasks of practical reason, and in the entire field of ethics. That is why the categorical imperative reads thus: 'So act that the maxim of your will could always hold at the same time as a principle in a giving of universal law.'[24]

VI

Our aim was to determine where Kant, despite the continuity of European metaphysics, departs fundamentally from the ethics of the Greeks in his moral philosophy. We can now answer this question: the difference comes from a new idea of freedom, and from the separation of freedom and truth it forcibly effects. The intelligible nature of freedom is defined by the fact that the human being, in so far as it is free, steps out of its existence in time. Only through this transcendence does it become autonomous; only thus can it constitute itself as a subject. But it can use the freedom thus acquired in the field of its existence. It can control the appearing nature as if it were not subject to heteronomy in it. Where it does so, the separation between theoretical and practical reason that Kant uncovered in the *Critique of Pure Reason* – the barrier of human finitude – is broken through. The science known as 'ethics' would have to solve the problems resulting from this new historical situation, but then it cannot be 'science' in the accepted sense of the word, for Kant showed that the knowledge of human freedom exceeds the thought forms of theoretical cognition. And so the doctrine of what should be remains powerless before that which is, and is produced by the arrogated sovereignty of humans. Therein lies the 'ambivalence' of Kant's moral philosophy: the dilemma in which we find ourselves results from an

[24] Kant, *Critique of Practical Reason*, p. 28.

understanding of the truth of which it is evident that it must destroy human existence in nature. We are forced to ask anew about the nature of truth. If we do not, we can no longer claim that ethical postulates are true.

In the European tradition, the word 'true' is used as an equivalent of the word 'right'. The usage of the word is ambiguous in itself: where it concerns propositions that hinge on the 'rightness' of facts, one understands truth as an agreement between our thoughts and the facts. If, on the other hand, one uses 'right' to indicate that acts of thought (propositions) agree with principles – whether those of logics, metaphysics or ethics – then one is dealing with a very different form of 'agreement'. Here, according to Kant, truth rests on that agreement of thought with itself whereby human thought qualifies as a 'subject'. Agreement with the identity of the subject is something different from agreement with facts. If the 'rightness' of ethics rests in its compliance with norms, and if this latter is attributed to the identity of the subject, the aforementioned dilemma results: that which is true is that which should be, but is not. As facts do not follow ethical norms, ethics is virtually defined by the circumstance that it does not agree with the facts and hence, by the standards of theoretical cognition, is untrue.

In the face of this dilemma, it can be instructive to consider that this use of the word 'right' does *not* correspond to what the Greeks meant by the word *orthos*. This word has three different meanings that were constantly confused with one another in later times:

1. *Orthos* really means 'upright'. In Greek literature, every form of delusion, deception or error is referred to with falling or toppling metaphors. By contrast, those who recognize the truth are 'upright'. The *orthos logos* of Stoicism is the upright logos. It is upright because it is in agreement with itself (*homologia*). One can show that Kant's concept of reason and the categorical imperative come from the Stoic tradition of upright thought, *recta ratio*.
2. A different metaphor for the recognition of truth is the straight path, as opposed to the crooked path or diversion. This is the decisive meaning of the word for Plato and Aristotle; the *orthos logos* of Aristotle's ethics is the straight path of the cognitive faculty.
3. If one wishes to build in an upright or straight fashion, one needs a yardstick. The Greek word for this is *kanon*, the Latin *norma*. The word 'yardstick' is also used metaphorically, in which case it refers to the 'rule' that human actions should follow, just as beams are set up using a yardstick. Polykleitos called his famous statue *kanon* because it was designed as a model for the 'right proportions' of the human body. Here, the meaning of *orthos* is akin to 'in proportion'.

None of these meanings of *orthos* have anything to do with the concept of truth as agreement between perception and fact. All three refer not so much to propositions or judgements as to the facts themselves; this also applies to the *kanon* of Polykleitus. Only the post-Aristotelian understanding of logic explains why the 'rightness' of a proposition or judgement was considered a *kanon* for all rightness as such. As propositional logic dictates that the *kanon* can only be true if it has the logical form of a sentence, the model for what is in proportion is remoulded into a proposition [*Satz*] that, in order that one can postulate its 'canonical validity', is termed a 'principle' [*Grundsatz*]. The transformation of the *kanon* into a principle then allows the other meanings of *orthos* to be subordinated to the logic-based understanding of truth, and reinterpreted in such a way that the view of truth expressed in them is buried. Thus ethics becomes a 'normative science' that remains in limbo because the truth of its 'principles' cannot be shown based on the understanding of truth it adopts – for these are defined by the fact that they do *not* correspond to reality. The predominance of the logic-based understanding of truth, especially in the Anglo-Saxon world, has resulted in the name 'ethics' referring almost entirely to the examination of moral judgements. The originary problem of ethics, the simple question of how to realize good through action, is then no longer posed.

The notion that ethics must be constructed as a science that derives the rules of practical behaviour from highest principles under changing circumstances follows the model of the demonstrative sciences presented by Aristotle in the *Posterior Analytics*. One must examine the reception history of this model to understand how anyone could decide to start from the highest propositions and not derive other propositions from them, like Aristotle, but rather something heterogeneous, namely courses of action. But even if this were possible, one would have to follow the example of Aristotle and ask under what conditions, and in what sense, one can state that the highest principles are true. Aristotle answers this question for the precisely circumscribed area of the demonstrative sciences through his 'first philosophy', which is not itself built on the model of a demonstrative science. If ethics is a 'normative science', if it is meant to proceed from highest principles, it must not be allowed to avoid having to substantiate the truth of its highest principles. But this cannot be resolved as long as one neglects the problem of how the truth of the highest principles is connected to the truth of the circumstances in whose midst we must lead our lives.

Kant saw through this problem very clearly. For him, the principle of ethics is not a 'Grundsatz'; rather, its 'logical' form is the imperative. In his moral philosophy, this form results from the structure of the faculty of desire; it is the faculty of desire that defines life. Kant discovered that ethics can only be true if its thought form is isomorphic to the structure of life as such. One can only interpret the 'logical' form of the imperative by examining how we are to envisage the edifice of nature if it is to be possible that living beings in nature act according to reason, that is, out of freedom. The analytic of pure practical reason methodologically presupposes Kant's teleology of nature, which he conceived in the methodology of the *Critique of Pure Reason* and elaborated in the *Critique of the Power of Judgement*. Thus the categorical imperative shows, as if under a magnifying glass, the unity of physics, ethics and logic that is noted at the start of the *Groundwork of the Metaphysics of Morals*.

In the period after Kant, philosophical ethics could no longer be built up as a doctrine based on axiomatic principles, for Kant had proved that there can be regulative *concepts*, but not regulative *principles*. Regulative concepts were introduced by Kant as his interpretation of Platonic ideas. The philosophy of the nineteenth century (in Germany since Lotze) interpreted the ideas as 'values'. Ethics became 'philosophy of values' (axiology). Under the influence of neo-Kantianism, this approach established itself in Britain (Moore) and the USA (Ralph Barton Perry, *General Theory of Value* [1926]; John Dewey, *Theory of Valuation* [1939]. This development reaches its apex in the historically representative ethics of Nicolai Hartmann. What is a value?

Balzac describes the history of this concept – if such a thing can be called a 'concept' – as follows:

> The general upset of 1830 brought to birth, as everybody knows, a number of old ideas which clever speculators tried to pass off in new bodies. After 1830 ideas became values. A writer, too wise to publish his writings, once remarked that 'more ideas are stolen than pocket-handkerchiefs.' Perhaps in course of time we may have an Exchange for thought; in fact, even now ideas, good or bad, have their consols, are bought up, imported, exported, sold, and quoted like stocks. If ideas are not on hand ready for sale, speculators try to pass off words in their stead, and actually live upon them as a bird lives on the seeds of his millet. Pray do not laugh; a word is worth quite as much as an idea in a land where the ticket on a sack is of more importance than the contents. [...] Having become an article of commerce, intellect and all its products must naturally obey the laws which bind other manufacturing interests.[25]

This chronicle of the genesis of a concept was written in 1833. Talk of 'moral and aesthetic values' only developed in France (according to the *Dictionnaire* by Robert [1976]) in the mid-nineteenth century. 'En morale, la notion de valeur a remplacé celle d'obéissance à une loi révélée' (Ribet).[26]

The difference between an 'idea' and a 'value' is easy to determine. Ideas are absolutely true, indeed the source of all truth. According to Plato, they are 'models that stand in nature'. Values, on the other hand, are what remains of ideas if one no longer believes in absolute truth. With the sharp eye of a conservative social critic, Balzac recognized what it meant that a metaphor from the language of the stock market was chosen to refer to something good or beautiful. 'Value' is the symbolic concept of capitalism. Metaphors have their own immanent logic: anyone who says 'value' implicitly has the relativity of all values and their inevitable devaluation in mind too. Furthermore, this word reflects the sovereignty of humans, who now believed themselves in control of ideas: values are produced, and can be manipulated. In political ideologies they are 'fixed'. According to Nietzsche, the principle of fixing values is the will to power. Economics and, in its service, politics thus enter ethics in unexpected fashion through the backdoor of 'values'. One should not speak of an 'ethics of business' without reflecting on how ethics, in being made a theory of value, subordinated itself to the primacy of economics, and to what extent the propagation of values – across the entire political spectrum – always serves economic interests too.

The ethics of the nineteenth century belongs in the history of the selling out of metaphysics. Today, no one dares any more to present, under the title of 'ethics', the matter itself – namely, a theory of human behaviours that demonstrates what good means for humans as such, and what it means in the context of ecological,

---

[25] Honoré de Balzac, 'The Illustrious Gaudissart', in *La Comédie Humaine of Honoré de Balzac*, trans. Katharine Prescott Wormeley (Boston: Hardy, Pratt & Company, 1900), p. 380.

[26] 'In morality, the notion of value has replaced that of obedience to a revealed law' (trans.).

religious, cultural, political, economic and social conditions under which they live. Ethics equivocates and becomes a theory of forming ethical judgements. Speaking about human behaviour conceals the actual behaviour and its problems; it conceals the question of the horizons in which it must develop and preserve itself. Nietzsche called this way of thinking 'nihilism'; but we cannot do away with the great questions of metaphysics by no longer adopting the answers it provided. The ecological crisis demonstrates that humans relinquish what they consider good if the human power to control exceeds the boundaries imposed by nature. Thus the connection between ethics and physics manifests itself. The destruction and dissolution of humanity taking place on all continents in the twentieth century shows us what happens to humans when they forget to ask what the truth is. The humans of the twentieth century no longer see themselves as 'the beings that possess the logos', that is, the beings that see the truth. Until Hegel, the determination of the nature of humans based on their recognition of the truth was the content of the lost philosophical discipline known as 'logic'. This gave rise to the inseparability of ethics and logic. The dimensions one must explore in order to understand what is good for humans have not been shifted by the crisis of European metaphysics. It is the fate of humans that they can only exist when they are capable of living in these dimensions – in other words, when they have an ethics that does not break out of their natural disposition. The problems discussed in philosophy and elsewhere today under the name of 'ethics' are of a largely ephemeral character, and simply reflect the discomfort of a civilization that is beginning to understand that the human domination of nature is being thwarted by the nature of things. Ethical reflection as practised today is of no significance for the future of humanity. The effort of thought that would be required of us in order to rediscover the true dimensions of human existence takes place where it becomes visible, under different names in the present day. People have forgotten that it was once called 'ethics'. As long as this is the case, the only observation we can make is that there is no such thing as an ethics that does justice to the philosophical concept of this science.

Part III
Time and History

Chapter 9
Time and Modalities (1971)

Everything that is, is in time. Everything that was, was in time. Everything that will be, will be in time.[1] The verb forms 'is', 'was' and 'will be' indicate different temporal 'modes'. Whatever the meaning of the term 'modes', the only way for us to distinguish between present, past and future is by simultaneously assuming that time is a unity *in* this difference. In this sense, the *unity* of time is the universal horizon for everything of which we can say that it is, that it was and that it will be. Hence our cognition of everything that is – in the three temporal modes – is always determined by the preconception we have, however unarticulated it may be, of the unity of time. Every object, every structure, every relation, every law that we cognize or believe we cognize, is at once a mirror of that preconception of the unity of time which makes human cognition possible to begin with. In this sense, we can say that what we call 'the truth' is the appearance of unity in time that is reflected in each individual cognition.

European metaphysics from Parmenides to Hegel understood the unity of time as the identity of the self-same, unshakable and immutable eternity, abiding in itself, and interpreted eternity as the unchangeable presence of being. They had strong reasons for doing so. The eternal presence of being manifests itself in nature as the *continuity* of space and time. If a crack appeared in the continuity of space, the cosmos would burst apart. If the continuity of time were interrupted, there would no longer be time, but rather nothing. That is why the eternal and unchangeable sameness of being that manifests itself in the continuity of space and time seemed to be the foundation of the universe as such. The Greek philosophers gave the eternal presence of being, which holds together everything that is, in its sameness, the name 'God'. It was hence only the cognition of this One God, who manifests himself in everything that is, that made the cognition of all other phenomena possible. The philosophical cognition of God

[1] This text was first published as: "Die Zeit und die Modalitäten". In: H. P. Dürr (ed.), *Quanten und Felder: Physikalische und philosophische Betrachtungen zum 70. Geburtstag von Werner Heisenberg*. Braunschweig: Vieweg 1976, pp. 67–76. Also in: *Hier und Jetzt* Vol. I, pp. 362–374. The permission to republish this text was granted by Klett-Cotta.

E. Rudolph and J. Picht (eds.), *Georg Picht: A Pioneer in Philosophy, Politics and the Arts*, Pioneers in Arts, Humanities, Science, Engineering, Practice 19,
https://doi.org/10.1007/978-3-030-31790-4_9

is the ground of all other cognition. Theology is the primary ontology that grounds all possible knowledge. The metaphysical doctrine of God is at once the science of science, as it renders the conditions of possibility and the inner order of all human knowledge transparent by their origin.

If, in the continuity of time, the sameness of the eternal presence of being appears in the form of *immutability* then the so-called temporal modes cannot be derived from the nature of time itself. European thought in the age of metaphysics was therefore forced to depict the unity of time in a form that allowed the differences between temporal modes to be disregarded. Aristotle – contrary to the archaic Greek understanding of time – famously depicted the course of time using the image of a circular path. Today we can no longer grasp the notion that time moves backwards into itself, and one could show why we cannot. Certainly there are also speculations in modern astrophysics about the possibility that the state which has been hypothetically reconstructed as the initial state of our world returns at enormous intervals. This is a sign of the hidden power that ancient mythical images continue to exert on scientific thought in the twentieth century. It is not the result of a scientific analysis of the nature of time. One can at least understand how Aristotle arrived at his conception of the cyclical course of time. For it emerges with stringent consistency if one combines three axioms that were unshakable tenets of Greek thought from Parmenides to Aristotle: (1) the axiom of the immutable unity of being, (2) the axiom of the finitude of the world and (3) the axiom that the immutable unity of being manifests itself in the finitude of the world. The immutable unity of being grounds the constancy of the world in One eternal presence. The finitude of the world is the reason why the eternal presence cannot be directly present in it. The mediation between eternity (1st axiom) and finitude (2nd axiom) results from the concept of appearance (3rd axiom). The medium of appearance is time, which, understood as a continuum, is eternal presence, understood in its passing, is *mere* appearance. The temporal modes belong on the side of mere appearance. If the passing of time takes place within a finite world, and if the unity of being in its self-identical presence is simultaneously to manifest itself in this passing, then the ever-same relation of the path of its passing to the steadily self-identical calm of constancy can only be depicted with the image of a circle on which every point is equidistant from the centre. Only if the course of time does not move away from the centre of identity can the principle still apply that the presence of eternity appears in this course. Therefore, the image of the circular path results with rigorous consistency from the premises of the eternal presence of being and the finitude of the world. The idea of the eternal presence of identity develops with astonishing consistency from a conception of God that already began to develop in Homer and Hesiod. It makes the form of thought later known as 'philosophy' possible by 'presenting' the unity of time in a form that makes both the *continuity* of space and time and the *passing* of time transparent and comprehensible in the same context. All cognitions that became possible in the horizon of Greek thought reflect the idea that the unity of time is the eternal presence of identity.

Greek thought did not view finitude, in its relationship with the absolute, one-sidedly as a negative attribute of the worldliness of the world. For them, *peras* [finite] is a predicate of good and *apeiron* [infinite] a predicate of the bad. The cosmos is an image of God precisely because it is *finite*: it manifests the completed and *perfected* nature of the One God. From the advent of nominalism, the great idea of the infinitude of power, the *infinita potestas* of God, established itself in European thought, and Nicholas of Cusa took the step of concluding from the infinitude of God that the world was also infinite. This disables one of the axioms of the Greek doctrine of God: now one can no longer conclude from God's completeness that his image is finite. But Cusa clings to a different axiom: he states that God manifests himself in the world. Only thus does the infinitude of the world follow from the infinitude of God. If God is infinite, his omnipresence can reveal itself in a spatially and temporally infinite world in every place and at every point in time. The image of the circular path is not abandoned, for it remains the only form with which it is possible to show that every point on the path is equidistant from a calm centre. But an infinitely large circular path has the form of a straight line; thus the change in the conception of God leads to a transformation of the form in which the unity of time is imagined. Modern Age thought imagines time in the image of an infinite straight line. The unity of time is uniform one-dimensionality.

The conception of God changed in the *one* regard that the relationship between *peras* and *apeiron* was reversed: the nature of the God who is experienced in terms of his omnipotence no longer manifests itself in the complete finitude of the cosmos, but rather in the infinitude of the universe. The other axioms of the philosophical doctrine of God of the Greeks, and thus of metaphysics, yet are preserved. This is shown by the philosopher who thought through the hidden foundations and preconditions of modern thought, especially modern physics, with the greatest clarity: Kant. In the chapter of the *Critique of Pure Reason* entitled 'The Schematism of the Pure Concepts of the Understanding', he writes:

> Time itself does not elapse, but the existence of that which is changeable elapses in it. To time, therefore, which is itself unchangeable and lasting, there corresponds in appearance that which is unchangeable in existence, i.e., substance [...].[2]

This corresponds to what Kant says in the section 'First Analogy of Experience'

> The time, therefore, in which all change of appearances is to be thought, remains and does not change; since it is that in which succession or simultaneity can be represented only as determinations of it. Now time cannot be perceived by itself. Consequently it is in the objects of perception, i.e., the appearances, that the substratum must be encountered that represents time in general and in which all change or simultaneity can be perceived in apprehension through the relation of the appearances to it. However, the substratum of everything real, i.e., everything that belongs to the existence of things, is substance, of which everything that belongs to existence can be thought only as a determination.[3]

What understanding of time is expressed in these words?

1. 'Time remains and does not change'. 'Remain' means lasting presence. 'Not changing' means immutability. The nature of time, or more precisely, the unity

[2] Immanuel Kant, *Critique of Pure Reason*, Critique of Pure Reason, ed. and trans. Paul Guyer and Allen W. Wood (Cambridge University Press, 1998), p. 275.

[3] Ibid., p. 300.

of time, is self-identical, eternal presence. Although time is now thought as infinite, the predicates that define its nature are still the predicates of the God of Greek philosophy. It is not time itself that elapses, but everything that elapses does so *in* time. The immutability of time as such grounds the possibility of depicting the path of time's passing with the image of an infinite straight line that is uniform in itself and does not show the difference between past, present and future.

2. Time itself cannot be perceived; it attains representation through substance. What Kant means by 'substance' is not this or that substance determined in this or that way, but rather 'the real' of appearance, 'of which everything that belongs to existence can be thought only as a determination'. 'Substance' hence refers to the pure identity of being, which is presupposed whenever we present that which is in the categories of quantity, quality, relation and modality. This pure identity of being is understood as immutability, that is, as the negation of all change as such, and this pure negation of change represents to us the time that cannot itself be perceived. The German word *Vorstellung* [representation] is used in philosophy as the translation of the Latin term *re-praesentatio.* Amidst all change, the being of existence represents for us the immutability of the time that cannot itself be perceived. According to Kant, existence as *re-praesentatio* is appearance. The horizon of all possible appearance is that which is 'represented', i.e. made to appear, in every appearance: the unity of time. But the unity of time can only appear in the form of *re-praesentatio* if its concealed nature is pure *presence.* This presence is lasting and immutable. So for Kant too, the ground of the concept of time is the constant presence of eternity. This constant presence – the God of Greek philosophy – cannot itself be perceived; indeed, as Kant shows in the 'Transcendental Dialectic', it cannot even be thought by the finite understanding. It *appears* to us only in the mirror of substance. Time, then, has unconditional primacy over being. In Greek philosophy, by contrast, being had primacy over time. The primacy of time over being follows from the fact that modern thought inverts the relationship between finitude and infinitude while retaining all other predicates of the God of Greek philosophy.

Kant's doctrine of time is not the product of unbridled speculation. He arrives at it through transcendental reflection on the conditions of possibility for classical physics. It is a doctrine of the hidden implications of this physics. Kant reveals the preconditions that must be assumed in order to justify measuring all elapsing of manifestations *in* time with a parameter defined as a fixed straight line. As Kant shows, this image of the passing of manifestations in time cannot be empirically verified, for every possible empiricism already presupposes a preconception of the unity of time. On the contrary: the preconception of time underlying classical physics contradicts our direct experience because we always experience time in the difference of temporal modes, whereas the image of an infinite straight line does not permit any structural difference between past, present and future. For this reason, the construction of the parameter of time cannot show that time has a direction and is irreversible. It makes no difference in which direction one passes along the straight line. This is expressed in the fact that the laws of classical mechanics do not show the irreversibility of time either. According to its equations, every movement could also run in reverse like a film played backwards.

But the depiction of all movements in time with the image of an infinite straight line had a further, very far-reaching consequence. If one presupposes that past, present and future are isomorphic, then what applies to the past also applies to the future: it is fixed. That is why the underlying conception time on which classical physics was based paved the way for strict determinism, that is, the doctrine that all future events are clearly determined according to unchangeable laws, and that it is only due to the limitations of the human cognitive faculty if we are unable to predict even today, with absolute certainty, all events that will occur at any given point in the future. This follows from the premise that all processes in time move in a single direction and that this direction always remains the same. Only if one combines the principle of causality with the principle of the rectilinearity of time, that is, with the elimination of the difference between past, present and future, and if one also presupposes that the world is a closed system, does the unambiguous determination of all processes at all times follow from this causality. Kant also analysed this connection with total clarity; he notes that the understanding confers 'temporal order on the appearances and their existence by assigning to each of these, as a consequence, a place in time determined *a priori* in regard to the preceding appearances, without which it would not agree with time itself, which determines the position of all its parts *a priori*'.[4] This confirms the statement that our preconception of the unity of time determines the form in which phenomena, which are in time, appear to us. The depiction of the course of time with a rectilinear parameter forces a corresponding formulation of the concept of causality. This formulation of the concept of causality determines the form in which we objectify the objects of possible cognition. At the same time, this concept of causality asserts the irreconcilable opposition between nature and freedom, an opposition which then determines the form in which modern thought brings the hemisphere of spirit and history to representation. If we had reason to revise the conception of time in classical physics and its metaphysical foundations as uncovered by Kant, we would be forced to undertake a corresponding revision of all fundamental concepts in modern science. The physics of the twentieth century should also encourage philosophers to rethink the question of our preconception of the unity of time, which is not a question of physics but of philosophy.

The unsolved and ever-suppressed problem in the metaphysical understanding of time is the difference between the temporal modes. What is the term 'modes' even supposed to mean if the passing of time is depicted with the image of a uniform straight line, and if the isomorphy asserted in this image is the form in which we bring the continuity and thus the unity of time to representation? How can we reconcile our immediate experience of time in the difference between present, past and future with the equally immediate experience of the unity of time? How can we think the differences between the modes at all when we know that time is not a substrate that passes through different phases, but 'is' itself nothing but the difference between its

[4] Ibid., p. 310.

modes? This question is connected – not only in its wording – to another great crux of philosophy, namely the problem of so-called 'modalities'. We saw that everything that is, is in time. Another principle seems equally fundamental: everything that is, is either necessary or actual or possible. How are these two principles connected? Could it be that the difference between the three modalities has something to do with the difference between the three modes of time? Or how else could it be justified?

This question loses its meaning on the ground of classical physics, for in a thoroughly determined nature, the following principle would apply: everything that is, is necessary. Modern philosophy responded to this by pulling the distinction between the three modalities out of nature, as it were, and transposing it to the domain of thought. Not nature but the thinking subject is such that it is forced to distinguish between possibility, actuality and necessity. Everything is considered possible that *can* be thought without breaking the rules of thought. Everything is considered necessary that *must* be thought according to the rules of thought. But what would 'actual' then mean? Here thought is suddenly forced to break through its own sphere. Kant says: 'That which is connected with the material conditions of experience (of sensation) is *actual*.'[5] He even goes so far as to declare that 'the determination of my existence in time is possible only by means of the existence of actual things that I perceive outside myself'.[6] Thought alone will never establish what is actual and what is not; on the contrary, the actual is the 'condition' for thought, for without 'the determination of my existence in time', that is, without experience, thought is impossible. But this exit from the sphere of the subject has consequences, for we notice that the two other modalities relate to the modality of actuality: the possible is what *can* be actual, and the necessary is what *must* be actual. The modality of actuality takes precedence in the fabric of the three modalities. Hence the modalities cannot primarily be represented as modalities of thought, but are rather rooted in nature itself. Their primary locus is not in logic, as it is thought, but in ontology. But it was impossible to include the modalities there as long as physics was constrained by its presupposed conception of time to maintain that everything that is, is necessary. This led not only in philosophy but also in the positive sciences to an ever-proliferating confusion, for on the one hand, people felt compelled to classify events or states as possible or probable, yet on the other hand, they were thrown back to the position of viewing statements containing the words 'possible' or 'probable' not as statements about the states or events themselves, but as statements about our cognition of these states or events.

The confusion becomes even greater if one realizes that we think not only *in* the modalities, but that we can cognize the modalities ourselves. It is *possible* for us to think possibility, actuality and necessity. Whenever we think, we *actually* think possibility, actuality and necessity. Indeed, it is even *necessary* for us to think possibility, actuality and necessity. So we encounter the three modalities here at a second order, in such a way that each of the second-order modalities can be placed above all three first-order modalities. But the distinction between first-order and

[5] Ibid., p. 321.

[6] Ibid., p. 327.

second-order modalities only becomes transparent when one understands that the terms 'possible', 'actual' and 'necessary' mean something different at the second order and the first. At the first order, the 'necessary' is that which must *be*, and at the second order it is that which must be *true*. Accordingly, what can *be* differs from what can be *true*, and what *is* differs from what is *true*. Because the second-order modalities are superordinate to the first-order modalities, however, there is reason to suppose that we perceive everything that *is* in the modalities of the first order, since the second-order modalities offer us access to what is *true*. The triad of modalities would then be grounded in the structure of truth. Assuming the triad of modalities had something to do with the triad of temporal modes, this could illuminate the mysterious relationship between the two 'orders' of modalities. If the three first-order modalities relate to being in the three temporal modes, every cognition that extends across all three modalities, and hence all three modes, refers to the *unity* of time. The second-order modalities would thus be different modifications of the appearance of the *unity* of time. If it is simultaneously true that we see in the second-order modalities not only what *is*, but what is *true*, this confirms the statement that truth is the appearance of the unity of time. But what does it mean that the difference between the modalities enters the very structure of the truth? It can only mean that the truth itself is not 'supratemporal' but 'temporal'. As the appearance of the unity of *time*, truth can only come to light in the difference between the temporal modes. We are thus taken back to the connection between the modalities and the structure of time even when we attempt to grasp the meaning of the modalities from the perspective of the structure of 'thought'. Yet are we justified in assuming that the difference between the modalities is based on the difference between the temporal modes?

One can reach an understanding of the connection between the temporal modes and the modalities very simply. (A comprehensive analysis of the highly complex structures that can be identified in this realm of possible knowledge cannot be carried out here.) What is constitutive for time as such and every experience of time is precisely the attribute that is *not* depicted by the image of the infinite straight line: its irreversibility. Time has a direction, and it is impossible to pass through it in reverse. Because time as such has a direction, everything that is in time must have the same direction; it is this direction that determines its possibilities. This applies as much to nature as to thought. If one asks what it means for time to have a direction, one first of all encounters a basic principle that can perhaps be considered the fundamental axiom of our entire cognition of nature. This principle is the following: it is impossible to undo something once it has happened. Certainly an existing state can change – indeed, it must change because it is in time. But each new state follows from the previous one; had the previous state not been, the new state could not be either. The new state thus contains everything the previous state communicated to it. The previous state is preserved in the new state; it is present in it as a communicated state. One can therefore encapsulate the statement that it is impossible to undo something once it has happened in this simple maxim: nothing past passes away. The principle of causality rests on this. It does not mean that the space of possibility for future states is *totally* determined; it simply means that the space of possibilities is limited by the law that each new state must contain the presence of all previous states that is communicated to it.

We have now made use of the concept of possibility. Only what can be in the future is possible; for what is present now is termed actual, and all that has passed must, because nothing past can pass away, be considered necessary. We refer to the space between what must necessarily be and what cannot possibly be as possible. Every process in time moves into this open space. Time is thus built asymmetrically; its structure is such that everything past is fixed by it and everything future is kept open within the boundaries resulting from the fact that the continued existence of the past must also persist in the future. Time is directional because of the asymmetry between open possibility and a past that necessarily continues in the future. The difference between the temporal modes manifests itself in this asymmetry.

Everything that can be actual is possible; everything that must be actual is necessary. Accordingly, we call what can yet be present 'future' and what was once present 'past'. The position of the modality of actuality corresponds to the position of the mode of the present. Like actuality in the fabric of modalities, the present takes precedence in the fabric of modes: past and future relate to (possible) present, necessity and possibility to (possible) actuality. How do we view 'present', how do we understand 'actuality', if we relate these two terms to each other? It is clear that we then cannot see 'present' as one particular point on a linear temporal scale. Then we are forced to consider other dimensions of the presentness of the present. According to Kant, as we saw, 'That which is connected with the material conditions of experience (of sensation) is *actual*'. We assure ourselves of the materiality of an experience by showing that what we claim is actual is present in a particular place or was present at a particular point in time. Actuality is defined by presentness *within a network of communications*. The word 'present' has no possible meaning for us outside of the multi-dimensionality of the frame of reference in which actuality appears to us. Rather, this multi-dimensionality itself is what defines the nature of the present. Present is the multi-dimensional horizon within which the actual appears.

Actuality is direct or mediated presence. The meaning of the modality of actuality can only be described in the universal interdependence of a communication context. Within this communication context, everything that we call actual, and thus present, has the basic character of manifesting itself. We refer to that which manifests itself – in whatever way – as a 'phenomenon'. In this sense, present would be the horizon of the phenomenality of phenomena within a universal communicative context. We call everything that manifests itself within this horizon actual. The other two modalities would have to be interpreted in a new analysis with reference to the concept of actuality thus attained.

The physics of the Modern Age does not refer to the communication context of what is simultaneously present as time; it calls it 'space'. In space different things exist alongside one another, in time they exist after one another. But 'alongside one another' always means simultaneously present and present within a communication context, for things that cannot communicate with one another cannot be said to exist within the same space. But if the space is constituted by simultaneity, its phenomenality is constituted by time, or more precisely: by a particular temporal mode, namely the present. We can then no longer distinguish between space and time. Instead, we must say that the multi-dimensional presentness of present manifests itself in so-called space.

If space is time, one can understand how spatial coordinates are connected to the temporal coordinate which is only superficially added in classical physics. The statement 'everything that is, is in time' then gains its demonstrable meaning. But then the direction of time can no longer be depicted as a linear curve; then time is a multi-dimensional, open fabric with mobile parameters, and every temporal determination requires the stipulation of a particular frame of reference within this fabric that is assigned to a particular 'location' and focuses on a particular cross-section of the multi-dimensionality of time. From a different 'location' one would reach a different temporal determination through a different frame of reference. The relation between the possible 'locations', however, can only be determined in the context of the temporal modes and the modalities; it is itself temporal. Time is therefore the universal horizon of the phenomenality of phenomena as such within the space of its modalities, which are grounded by the temporal modes.

Truth is the appearance of unity in time. How is the cognition of the *unity* of time at all possible for human thought *in* time? We are forced to ask this question, for all human thought has the structure that it can only develop the historical manoeuvring space of its cognition through the question of its own possibility. But do we understand this question? What does 'possible' mean here? In what sense are we using the word 'possibility' here?

At the first 'order', 'possible' means that which can *be*. It is in this sense that the term was assigned to the temporal mode of the future. At the second 'order', 'possible' means that which can be *true*. In this understanding, possibility, like the other two 'modalities', was placed above the three temporal modes. Now the three modalities related to different modifications of the appearance of the *unity* of time. But if we ask how the cognition of the unity of time itself can even be 'possible' for an actual entity situated in its own present, we transcend the second order too. The 'possibility' of the appearance of the unity of time as such cannot be equated with a particular modification of this appearance. This new concept of possibility differs from the other two 'orders' in that it bursts open the fabric of the three modalities. Even if it is 'possible' for us to cognize the unity of time, we can never say that it is actual, let alone necessary. It is, to use a Kantian term, only possible in the form of an 'infinite task'. How should we interpret this concept of 'possibility'?

Since Kant, the question of the conditions of possibility for cognition has been described with the term 'transcendental'. Kant still interprets this term within the horizon of the immutable unity of the faculty of reason, that is, the horizon of the timeless presence of identity. That is why his 'transcendental philosophy' is metaphysics – a metaphysics of finite reason. But the terms 'infinite task' and 'possibility' can only be understood within the horizon of the future; they resist any attempt to bend them back into an eternal present. But here the futurity of the future shows itself to us in a form that – analogously to the new concept of possibility – can no longer be integrated into the fabric of the three temporal modes. So what is the word 'future' supposed to mean?

The structure of the three temporal modes was, as became apparent, held together by the primacy of one particular 'mode', namely the present. This primacy was based on the fact that the present proved to be the universal horizon for the phenomenality of all possible phenomena. Thus the present grounded the actuality of what we call actual. We therefore call the form of time that unfolds in the difference between the three modes '*phenomenal*' time. In the possibility of the cognition of the unity of time, we also encounter the future, and thus time – but in a different form that we cannot explicate here. We call it '*transcendental*' time. This should not be taken to mean that there are two different times here; they are two forms of one and the same time that cannot be reduced to each other. We can only cognize the unity of phenomenal time because we always simultaneously experience transcendental time, and because the unity of phenomenal time stands apart from transcendental time in our experience of time. For this reason, the following statement can be made: all human thought operates in the difference between phenomenal and transcendental time.

How is the experience of transcendental time itself 'possible'? In the age of metaphysics, all cognition was a mirror of the unity of phenomenal time that came to light in the idea of eternal presence. For the Greek philosophers, God manifested himself in this idea. Does transcendental experience of time also become possible through a revelation of God? It follows from the transcendental concept of future that this is a question one can only understand for as long as one is determined to keep it open.

Chapter 10
The Historical Nature of the Human Being (1976)

The theory of the nature of humans has, under the name 'anthropology', become a fashionable science.[1] The market is flooded with biological, medical, psychological, pedagogical, philosophical and theological anthropologies, cultural anthropology and social anthropology.[2] The term 'anthropology' risks becoming a mere advertisement, drained of all meaning. It serves to lend a scientific veneer to pre-scientific convictions and ideologies; critical anthropology that attempts to reveal this has little effect, or becomes entangled in the very mentalities it attempts to eliminate. Few people realize what they are actually saying when they use the word 'anthropology'. We will therefore begin by asking: what does 'anthropology' mean?

Zoology is the science of animals, sociology is the science of society, psychology is the science of the soul. Accordingly, anthropology is considered the science of humans. What is a human being? According to Aristotle, it is the *zoon*, the living being that possesses the logos. If a human develops a zoology, they will appear twice there: as a living being, they are an object of this science; that is why there is a 'zoology of humans'[3]. Because they possess the logos, they develop this science. They are at once its object and its subject.

Sociology is the science of society: the Greek word for society is *polis*. According to Aristotle, the human being is not only the creature that possesses the logos; it is also the *zoon politikon*, the creature that exists in a society. These two definitions are not superficially added together, but rather condition each other: humans live in society because they have the logos, and they have the logos because they live in society. Sociology thus finds itself in the same situation as zoology: society is the

---

[1] This text was first published as: "Die geschichtliche Natur des Menschen". In*: Mensch und Technik – Leben und Tod*. epd-Dokumentation 23 (1976). pp. 5–18. Also in: *Hier und Jetzt* Vol. I, pp. 165–181. The permission to republish this text here was granted by Klett-Cotta.

[2] This essay gives a summary and overview of the preparatory work for my three essays in Constanze Eisenbart (ed.), *Humanökologie und Frieden* (Stuttgart: Klett-Cotta, 1979); the ending presents the conclusions from the book *Theologie – was ist das?* (Stuttgart: Kreuz-Verlag, 1977).

[3] Joachim Illies, *Zoologie des Menschen: Entwurf einer Anthropologie* (München: dtv, 1976).

E. Rudolph and J. Picht (eds.), *Georg Picht: A Pioneer in Philosophy, Politics and the Arts*, Pioneers in Arts, Humanities, Science, Engineering, Practice 19,
https://doi.org/10.1007/978-3-030-31790-4_10

subject area of a science that itself emerges from this society. The same society that is an object of analysis determines the forms of thought and the methods of analysis from the background.

It should be self-evident that psychology, in so far as it is knowledge, reflects the states of consciousness and attitudes of the same soul that it studies. But the modern psychology modelled on natural science has repressed this fact so vehemently that it imagines it is able, through a soulless insight, to grasp the soul as it is. What it represses is projected as the mirror image of its unconscious, and then appears as the object it gazes upon. Such is the mysterious nature of the relationship between science and its subject area. According to the teachings of Plato and Aristotle, humans can only have the logos and be political because they are ensouled. Psychology too, whose history begins with Aristotle's work *De Anima*, is therefore based on the Aristotelian definition of the human being. But the Greeks were careful not to limit the use of the word 'soul' to humans; *psyche* refers to the inner structure of all organisms, that is, the 'life' that distinguishes them from inorganic nature.

Because zoology, sociology and psychology are all rooted in one and the same definition of the human being, anthropology constitutes their axis. It would thus be the science with the function of elucidating the connections between nature, society, soul and knowledge in humans and the ways in which they are intertwined. These different sciences, as '-ologies', intersect in the logos, on whose structures they are modelled. Hence the age of metaphysics, from Aristotle to Hegel, did not interpret the inner nature, the essence of humans in terms of the manifold ways in which the individuals we call 'humans' appear to us in nature, in society and in their own self-awareness. Rather, it was thought that the essence of humans distinguishing them from other creatures lay in those forms of knowledge that enabled them to conceive a science of nature, of animals, of society and the soul, and in all these also of themselves. The essence of humans was sought not in the object spheres of science but in science itself.

I

As in zoology, sociology and psychology, the human being appears twice in anthropology: as the *anthropos* that is its object and as the logos that determines the perspectives from which this object is viewed. Contemporary anthropology is primarily interested in the *anthropos*; it has largely repressed and forgotten the 2,500-year investigation of the logos. The age of metaphysics concentrated on the logos and related all manifestations of the *anthropos* to this logos.

In the meantime, however, the process of history has taken us to a point where we should recognize that the true essence of the 'nature' of humans emerges neither in the *anthropos* nor the logos. It lies hidden in the inaudible hyphen that binds these two parts of the word 'anthropology' in a way that has not yet been illuminated.

In order to approach an understanding of this hyphen, we will ask in the first part how anthropology understands the *anthropos*, and in the second part, how it understands the logos and how each is reflected in the other. This will allow us, in the third part, to proceed to the heart of the mystery indicated by the hyphen.

According to a legend that is currently being propagated even by respected exponents of the field of 'philosophy', philosophical anthropology 'as an independent discipline is only a child of our century';[4] its birth, as one reads time and again, can be dated to 1928, when *Man's Place in Nature* by Max Scheler and *The Stages of the Organic and Man* by Helmuth Plessner were published. The third work that represents this discipline's way of thinking is Arnold Gehlen's book *Man*,[5] whose subtitle *His Nature and Place in the World* refers back to Scheler. These books do not present a new beginning, however; they echo the programme of Feuerbach, who demanded the replacement of theology with anthropology over a hundred years ago.

But the secular, anti-theological anthropology of the Modern Age has much older origins: it began with Descartes's *Treatise on Man* and Hobbes's *De homine*. Here, for the first time, all functions of the extended body we call 'human', including sensory perception, imagination and memory, were explained in purely mechanistic terms. As we know, the scientific anthropology of the twentieth century has remained true to this mechanistic programme in its main currents to this day. Thus the great shift in the view of the human being took place not in the twentieth century, but already in the seventeenth. While the functions of the brain, according to this interpretation, obey the laws of mechanics, the thoughts obey those of logic. They are not extended and cannot be quantified. Descartes acted on this with a rigorous separation of human consciousness, the *res cogitans*, from the physical object of the extended body. When he explores the *res cogitans* he does not examine the functions of the brain, but rather seeks to ascertain how the thoughts produced by this brain can be true and lay the foundation for a rational knowledge of the entire universe. For this he requires the help of metaphysics, while the same question leads Hobbes, who rejects this way out, to adopt a radical scepticism.

Already before the end of the seventeenth century, however, a great mathematician and physicist – Leibniz – initiated a counter-movement to Descartes's mechanistic view of nature. Because the basic phenomenon of nature, life, cannot be explained in a mechanistic system, he did away with the dichotomies that had methodologically enabled the physical worldview: the opposition between matter and spirit, between inorganic and organic nature, between subject and object and between causalism and theology. Following on from Leibniz, Herder modified the separation of human and animal rooted in the Stoic-Christian tradition. In section 80 of the *Critique of the Power of Judgement*, sixty-nine years before Darwin, Kant proposed the idea that the creation of organized living beings in nature leads from raw matter via mosses and lichens, aquatic animals, swamp animals and land animals to humans, and that this process of development can be explained according to certain laws. Two years earlier, in the field of comparative anatomy, Goethe had torn down the last barrier between human and animal with the discovery of the incisive bone. German Idealism and Marx are also part of the history of this counter-movement to the mechanistic worldview. In

[4] Otto F. Bollnow in *Philosophische Anthropologie heute*, ed. Roman Roček and Oskar Schatz (Munich: Beck, 1974), p. 19.

[5] Arnold Gehlen, *Man: His Nature and Place in the World*, trans. trans. Clare McMillan and Karl Pillemer (New York: Columbia University Press, 1987).

1876, under the title *Human, All Too Human*, Nietzsche produced the first outline of an anthropology that, with a boldness that would never be equalled, followed on from Lamarck and Darwin to show the necessary implications of the principle of evolution for the understanding of humans, their morals, their religion and their culture. In so doing, he anticipated the central ideas of psychoanalysis and behavioural research without succumbing to the hasty conclusions of their methodology. It was from Nietzsche that Gehlen took his basic thesis that the human being is the 'not-yet-determined animal'. Ruth Benedict also refers to Nietzsche, as well as Dilthey, for the methodology of her cultural anthropology; modern social anthropology is unthinkable without Marx. In its methodology, the anthropology of our time has fallen far below the level already reached by these great thinkers, because it is dominated by the tendency to ignore its own historical origins. And it is hardly fitting for the science of evolution to deny its own evolution.

Once the theory of evolution had established itself, a strictly mechanistic explanation of nature should have become impossible, for the latter is only consistent when it asserts that every event in nature is determined unambiguously by constant laws. This presupposes that the universe is a closed system, that the passing of time is correctly represented by a straight line, and that all processes in nature obey logic. These assumptions cannot be empirically verified; on the contrary, this logic is used to define what is permitted by natural science as 'empiricism' to the exclusion of all other forms of experience. The parameters of the mechanistic worldview, as Kant already showed, are projections of the logos. The *anthropos* that appears on the screen is the product of thought operations behind the back of the observer. Hence Plato's image of the bound prisoners in the cave who can only see the shadows cast on the wall facing them is an exact description of the perspective from which mechanistic natural science envisages the species called 'anthropos'. And one should not be fooled by the refinement of scientific methods; there is a direct path leading from the automata of Descartes to the hydraulic models of drives used by Freud and Konrad Lorenz.

The singular 'human' is already a projection of the logos; for it follows from the theory of evolution that there are no self-identical and unchanging species in nature. Today we are as much in the 'transitional zone between animal and human' as was *Homo pekinensis*. For the transition from animal to human still takes place in the development of each individual, and if we open our eyes, we can see how often it fails. The fact that our brain has a considerable weight (and that its performance could possibly be enhanced by implanted computers) does not stabilize the achievements of the species; the range of possible failures increases in proportion to our chances. The word 'human' refers not to an object that exists, but to a constellation of possibilities in nature whose realization always remains uncertain. The search for 'anthropological constants' to define the human being as a member of a species follows an orthodox metaphysical thought model. But Kant already demonstrated two hundred years ago that we are falling prey to a transcendental semblance if we substitute real essences for concepts. Thus, upon closer inspection, the object of the science known as 'anthropology' dissolves before our eyes. The human being it deals with does not exist.

This result contradicts a conviction that consistently dominates positivist anthropology. In contrast to the metaphysical speculations in which the self-conception of humans had supposedly lost itself before the triumph of the natural sciences, it is driven by the ambition to be an empirical and exact factual science. According to the programme of Auguste Comte, which is already rather stale, the age of positivism will put an end to the dark age of metaphysics through a new Enlightenment. What is the nature of this empiricism and its facts?

There cannot be any human experiences of periods in which no humans were alive. The science of human evolution owes its success to a method that is only very indirectly connected to empiricism. This can be demonstrated with an example: on 9 March 1976, Richard Leakey announced that he had found a skull in Kenya which proved that members of the species *Homo erectus* had already lived there one and a half million years ago. This is considered empirical knowledge; we believe the photograph that the mineral-encrusted bone genuinely exists.[6] So how is it that Leakey reconstructs half a million years of human prehistory from this one skull? For this purpose, he requires a combination of insights from numerous specialized sciences whose results have been conveyed to him through tradition. The number of intellectual operations required to arrive at these results is inestimable, and impossible for any living human to reconstruct. We believe them because they have proved themselves in the tradition of the sciences and because we have no reason to doubt that each of these operations is fundamentally verifiable. In addition, he used devices that in turn contain endless amounts of information and intellectual operations. With these tools, he subjects the skull to a series of tests that allow him to observe certain things – not in the skull, but in the devices. From those he concludes that the *Homo erectus* to whom the skull belonged lived one and a half million years ago. The result of this conclusion based on belief and tradition is then presented as an empirically proven fact.

I have no intention of casting doubt on the truth of Leakey's discovery. I am simply noting that only a very small part of it rests on empirical observations, while the far larger part is based on logical operations. Logic is the instrument that enables modern science to break through the boundaries of human experience and reconstruct huge periods in the evolution of life long before the genesis of the human brain. At the same time, however, logic is meant to be derived from the brain's functions, that is, from evolution. How is it possible that logic already applied before there were people who had learnt to think logically? This was the old problem of metaphysics, which did not accept easy solutions as readily as modern natural science when it came to the rationality of humans.

If we question the absolute validity of logic and its implications, all reconstructions of evolutionary theory collapse. But trust in the absolute truth of logical and mathematical theorems is not gained through empiricism; it is a residue of metaphysics. Thus modern natural science is applied metaphysics. The only difference between them is that natural science does not investigate the grounds for its dogmatic faith in the truth of logic and mathematics, whereas metaphysics viewed the question

[6] *International Herald Tribune*, 10 March 1976.

of whether this is not a mere preconception as the greatest problem facing the human spirit.

The theory of evolution is a great discovery, though there will undoubtedly be changes in its interpretation. But we would be deceiving ourselves if we took it for an empirical factual science. What does it actually teach? It shows that we can still only understand ourselves and the forms of our experience by overstepping the limits of that experience. We explain the nature of the contemporary human being by sending ourselves back along the straight road of linear time into epochs hundreds or thousands of millions of years before the inception of its species, so that we can return from there to ourselves.

In search of the truth of the logos, the age of metaphysics overstepped the boundaries of human experience in a vertical direction, towards a justification of a priori knowledge. The anthropology of our century prefers the horizontal direction – it follows the axis of linear time. But in this dimension too, knowledge that goes beyond the boundaries of human experience must be termed transcendent knowledge. It is evidently a part of our life that we must go beyond ourselves, into the infinite, in order to understand ourselves and thus exist. Whether we follow the horizontal or vertical in this makes less of a difference than people think.

Hence the result of this observation of modern anthropology is the following: it is only possible because the human being – if I may use the word naively in the singular once more – is a universal living being, that is, a living being that can only understand itself within the horizon of the universe. The openness to the world that is constitutive of the human being knows no boundaries, and its orientation towards the world defines its way of life. That is why humans have always required some form of anthropology in order to assert themselves in the world. What they 'are' on the basis of their understanding of themselves and the world can no longer be determined as an object, for world is not objectifiable either; but the understanding of the world becomes a part of life. Hence the Greeks distinguished between the form of life that must be produced time and again by humans – *bios* – and life understood as a functional form of the organism – *zoe*. To ancient Greek ears, 'biology' would quite simply mean ethics. All that remains of the *bios* on the screen of anthropology, however, are the X-ray images that allow us to look through it and identify the behavioural patterns of our animal ancestors or our reflex reactions to economic conditions. Cultural products and forms of social organization are considered explained when we see through the biological or economic mechanisms without which they could not have developed. This is analogous to attempting to explain the artistic gestalt of a picture from a chemical analysis of the colours and the mechanics of the brushstrokes. What makes the picture a picture, its gestalt, has disappeared. And so we can find a great deal that fascinates us in the anthropology of the twentieth century, but not ourselves.

II

The human being, as we saw, appears twice in anthropology. It is the *anthropos* that it views as its object; but it is also concealed behind the logos whose rules govern the conception of this science. Modern anthropology mostly contemplates the *anthropos*. Metaphysics strives to find the nature of the human being in its capacity to recognize the truth; its subject is therefore the logos. We will now ask how the human being appears when it makes use of the logos.

In its usage by Plato and Aristotle, the word 'logos' does not mean 'reason' or 'language'; rather, it refers to a particular form of statement. A logos is a statement that uncovers those structures of reality which are constant and untouched by the alternation of becoming and passing. Plato calls such structures 'ideas'.[7] An idea is not a thought, something contained by the consciousness, but rather a model of the reality in which it appears, a model that stands for itself in nature and endures. It is the pure gestalt of that which is. When we identify an animal as a horse, we establish that this animal shows traits corresponding to the model 'horse'. Every individual horse reproduces this model. In biology we refer to such models as genus and species. Species are not something thought up; Darwin did not write his famous book about the evolution of thoughts. But they cannot be sensually perceived; only the examples of the species can be sensually perceived. That is why an idea, according to Plato, is a sensually imperceptible structure in nature. When anthropology speaks of humans, it does not deal with the sensually perceptible individual specimens but with the species itself, and attempts to identify attributes that apply to the species as such, and hence for each of its specimens. In this sense, anthropology of all types and schools, but most of all the natural-scientific type, is an 'idealistic' theory.

Natural human understanding resists the notion that something can be real if it is not sensually perceptible. It is very peculiar that this prejudice, which was already fought by Plato, could persist into the twentieth century, even though modern natural science had long since emancipated itself from the sensual perception of humans. It replaces the human senses with machines that can no longer be considered 'organ extensions', but have been constructed according to rational conceptions completely independently of the performance of the senses. Through complicated mathematical operations, the signals of these devices are translated into information about objective facts. The non-sensual world thus disclosed is the 'nature' of modern natural science. This science could therefore be interpreted as an empirical confirmation of the Platonic theory of ideas. But Plato did not stop at the structures he calls 'ideas'. He discovered that purely mathematical figures, numbers and the relations between them can be found in the background of these as something like the structure of the structures. These too are revealed by the logos; they too are models in nature. He uses this to explain that natural phenomena are determined by mathematical laws. The logos, the revealing statement, is the instrument that allows us to recognize these structures. It is the organ of the spirit, and according to Plato, to be human means to understand the showing in the logos.

The prejudice that only what we perceive with our sensory organs can be *real* led, in the later development of European philosophy, to the projection of everything that cannot be sensually perceived back onto the interior of humans, onto the 'soul'. Because the Stoics defined the logos not as the statement but as the divine element in

[7] In order to avoid distorting Picht's meaning, what is commonly referred to in English as Plato's 'theory of forms' is here termed the 'theory of ideas', in keeping with the German term *Ideenlehre* (trans.).

the human soul, the word 'logos' now took on the meaning 'reason'. The statements in which we recognize the truth are interpreted as functions of reason, which then leads to the possibility of transposing the insights gained through these statements into reason. The ideas are no longer genuine ontological structures found in nature, but are reinterpreted as thoughts and become 'innate ideas'. One can retrace step by step how ontological structures recognized in the course of European intellectual history were projected back onto the interior of humans; this gradually developed what modern philosophy would ultimately call 'reason'.

This is how the *res cogitans* of Descartes, which we tend to refer to in Kantian language as the 'subject', constitutes itself. The boundary between subject and object is a reflection of the boundary between sensual perception and non-sensual knowledge. Because the laws of logic could no longer be interpreted as understood laws of being, however, only as laws of thought, it was no longer explicable why natural phenomena, degraded to mere objects, should obey the laws of logic and mathematics. The immense intellectual work of transcendental philosophy from Kant to Hegel was devoted to solving this problem. It failed because its purveyors failed to recognize that it was a pseudo-problem. They remained in thrall to those projections from which the subjectivity of reason emerges in the first place.

The natural sciences were not interested in thought, only in the objects. They turned their back on their own preconditions in nature, which were now viewed as mere acts of thought. That is why natural science only ever sees half of nature, its object side. It follows from this, however, that the congruence with nature of the models projected onto the object sphere can only be verified in secondary terms; one cannot explain through natural science why it is possible to alter nature through science, that is, through thoughts. The deformation of nature that must be considered a result of modern natural science stems from this misconception of the logos.

In the allegory of the cave, Plato showed that the human being must undergo a reversal of the entire soul in order to break away from sensual phenomena and set eyes on the horizon of non-sensual structures in nature. Following on from Plato, Hegel still speaks of a reversal of consciousness. Because the sphere of non-sensual cognition appears reflected back inside reason, however, Plato's reversal of the entire soul is transformed into the movement known in recent philosophy as 'reflection'. Knowledge of the truth becomes, as Kant puts it, the 'self-knowledge of reason'. This reason no longer circles within nature, but rather understands itself as autonomous and accordingly develops a praxis emancipated from nature. This praxis cannot be explained by the behavioural patterns of our animal ancestors. It is a product of reflection that is guided by rational schematicisms. The primary form of praxis in the scientific-technological world is therefore theory.

Reflection is a movement that is not carried out by the soul in its entirety. In the nineteenth and twentieth century in particular, it became an isolated operation of human rationality. This too has a history. In his book on the soul, Aristotle already separated the purely cognitive functions of sensual perception from the affects; the theory of the taming of affects through reason finds its place in the *Ethics*. Thus begins a destruction of the whole sphere of human sensuality. Stoicism radicalized the opposition of reason and affects; its doctrine of the soul, mediated by Christian asceticism, has defined the self-conception of Europeans to this day. This separation and negation of sensuality is also a consequence of the projection of ideas onto the interior of humans. The result of this projection, Kant's 'logical I', is not affected by the stirrings of the soul. The subject of rational science can neither suffer nor feel happiness; it knows neither death nor life. It is an isolated rationality that imposes its soullessness on the world that it dominates through the mechanisms it produces. This can also be demonstrated with reference to anthropology: it is the mirror image of a thinking that reflected itself out of its natural context in the search for its autonomy.

Hegel fell prey to an illusion when he taught that the process of European intellectual history obeyed an immanent logic. Today, evolutionary theory succumbs to the same illusion. If one assumes that the passage of time is adequately represented by a rectilinear parameter, the past necessarily becomes a one-way street leading directly to us. And because nothing that has happened can be undone, we confuse the irreversibility of all past events with inexorable laws governing them.

A more precise analysis of the historical period in which humans interpreted their nature from the perspective of the logos would show just how heterogeneous the influences determining the direction of the process actually were. Each of its stages was determined by very different actual historical conditions, economic relations, social structures, the political game of power and cultural forms. All this is intertwined, alongside numerous connections to the past, with the immanent advancement of scientific thought – which by no means, however, proceeds as logically at all times as it tells itself, but is also driven by human, all-too-human motives beneath the surface.

If our image of the *anthropos* is a projection of the logos, it now becomes clear that the logos, conversely, is nothing other than a backwards projection of the history of the *anthropos*. To recognize this, however, we must put ourselves in the position of the captive from the allegory of the cave who rises and brings about a turn of the entire soul. We are now in an interstice bridged by the invisible hyphen between the two parts of the word 'anthropology'. This is the place of the human being as it truly is, but as which it cannot appear either in the image of the *anthropos* or the reflection of the logos. This, as Plessner said in one of his final contributions to anthropology, is 'Homo absconditus'. I will consider this in a short third section.

III

In 1926, two years before the anthropologies of Max Scheler and Helmuth Plessner, Martin Heidegger published *Being and Time*. This book too, under the title 'Existential Analytic of Dasein', contains a theory of the human being. But it speaks neither of the *anthropos* nor the logos; the foundations of all previous forms of human self-conception are called into question. The critical philosophy of Kant had still dogmatically presupposed reason as the source of a priori knowledge, but Heidegger goes beyond Kant; he examines the preconception of the nature of being and truth that is already built into Kant's concept of reason. The 'destruction of the history of ontology' he carries out unsettles all the prior assumptions that modern science had left unquestioned, including its anthropologies. The critical breakthrough achieved by Heidegger followed from the experience of foundational crisis that had shaken even the most solidly established science, such as mathematics and physics, in the 1920s. Heidegger refers to this at the start of the work (§3). He showed that a knowledge of the human being in the form of anthropology is not possible. We will adopt this insight as our point of departure here, even if we are forced to take a different direction.

Information about a phenomenon can only be gained from the experience in which the phenomenon is given to us. The experience in which we understand ourselves as humans places us in a multi-dimensional space and discloses itself in a boundless multiplicity of forms that are not consistent, but rather incommensurable. We will not undertake an analysis of these forms of experience, but simply observe that the empiricism of the modern positive sciences appears within the rich spectrum of human experience as an isolated extreme that only became possible through the specific conditions that obtain in an industrial society.

Empiricism is what we call the forms of experience that provide us with unambiguous knowledge. Nothing justifies the assumption that natural phenomena are inherently unambiguous; for every phenomenon, as Leibniz already recognized, reflects an infinite number of relations within itself. Unambiguousness only comes about if one creates conditions under which particular classes of relations manifest themselves, while the rest are hidden. Empiricism informs us of clearly defined *aspects* of phenomena; it does not inform us about the phenomena themselves. In the multi-dimensional space of human experience, by contrast, phenomena appear in their undelimitable ambiguity. They are not stylized into objects, but show themselves in the intangibility that defines their character. The human being too, in its own experience, appears for itself as neither subject nor object. One must therefore push aside all conventional thinking if something of the reality of *Homo absconditus*, whose hiddenness forms part of its nature, is to enter our knowledge.

However incommensurable the forms of human experience may be, there is one thing that can be said of them all: every experience includes the experience that it is taking place in time. Time is the rarely-mentioned yet universally underlying content of every experience. Of course, the form in which we encounter time in the experiences accessible to us is not the same form presented in the natural and historical sciences. We experience it not as a one-dimensional linear progression but always in the difference between its modes: past, present, future. The image of the straight line does not represent the difference between these modes. That is why the view of time that dominates our sciences contradicts our experience; but empiricism presupposes this view. It must presuppose it in order to provide unambiguous information. If, on the other hand, one thinks through the difference between the modes of time, to which the difference between the modalities of being – necessity, reality, possibility – corresponds, one enters a multi-dimensional area that can no longer be represented in a clearly appreciable way.[8] The movement of thought must pass through all dimensions of human experience if it is to explore the sphere of time. To encompass everything that can appear to us in time, we use the opaque

[8] See 'Time and Modalities', pp. 151–160.

term 'the world'. We can therefore say that the open multi-dimensionality of time sketches out the measurements within which we can experience what is in the world.

Everything of which we can say that it *is*, is in time. Even the supposedly timeless structures of mathematics are in time; for the eternity that we claim for them is nothing other than the unity of time as it appears to us when we imagine it in the temporal mode of the present. But the human being not only exists in time; it also knows that it exists in time. Where does it gain this knowledge?

The oldest documents of *Homo sapiens* known to us are burial sites and their furnishings. We can therefore say: humans are creatures that know in advance that they will die. Their life in time is accompanied by the certainty of their future exit from time. This means that they are simultaneously in the midst of time and distanced from this time. The form of life that results from this self-sublating being-in-time forces humans to shape their *bios* as their history. But one cannot acquire knowledge of something without distance from it; hence humans know that they are in time because the anticipation of their exit from time distances them from time with every breath. This is how they experience the transparency of time and all that appears in time. This experience is the origin of all knowledge.

The human being's oft-cited openness to the world is already given with the experience of time as time. Time is the epitome of the universal medium: it cannot be restricted to a particular environment, but breaks through all barriers that humans artificially install time and again to stabilize their cultural and social systems. Because every human being experiences time, it is a universal creature – even in the most absurd forms, even in the mode of negation. It does not only live in its environment, but in the world.[9] That is why Leibniz described life as the tension between individuality and universality represented in every monad. But because this tension is experienced in time, and thus as history, it is clear why evolutionary theory must go back to the genesis of the planetary system and organic life on earth in order to make sense for itself, as far as possible, of the human as a historical creature.

Until now, the theory of evolution has interpreted its observations and data according to the linear schema of cause-effect schema of classical physics. Going against this model, Wolfgang Wieser[10] has put up for discussion the cybernetic 'paradigm' of a systems theory in which the evolution of living beings is described as an evolution of open systems. This new version of evolutionary theory has its unmistakable limits too, but it supplies this important theorem: 'The specific information from the environment is a functional characteristic of the "organism" system itself'.[11] We will apply this statement to the human being and free its content from the precise, but accordingly narrow terminology of cybernetics. Then it becomes this: the discovered being of the environment is constitutive of the being of humans themselves. This aligns with Heidegger's principle that the human being *is* the disclosedness of its world. If these statements are true then no organism, least of all the human

[9] See 'Philosophie und Völkerrecht', in: *Hier und Jetzt* Vol. I, pp. 57–115.

[10] See 'Die Korruption einer Wissenschaft: Zur Lage der Verhaltensforschung', in *Merkur* 329/330 (November 1975), p. 1022.

[11] Ibid, p. 919.

being, can be viewed as an object that exists, and therefore also exists in an environment. But neither is the human being an incorporeal *res cogitans* that views its environment from a timeless vantage point as a mere object sphere. Rather, the environment is represented in the organism and the organism in the environment. This also applies to the human being, except that the human being knows that it is in time, and can therefore see through this.

But evolution, as we know, never erases any of an organism's earlier stages completely; thus every manifestation of life has an inbuilt memory of long-vanished environments. The atavisms of animal and human behaviour make this clearly visible, and psychoanalysis has developed methods of anamnesis that bring to light the prehistory which lies hidden within us. The creature that knows it is in time can break up, as it were, the memories that are latent in it and make them speak. It understands that, in its present, it *is* the history that it left behind after an optical illusion.

As this history has passed through all spheres of nature as we know it, it is not a myth but rather clear-sighted experience that humans since Plato have considered themselves a microcosmos. We cannot, admittedly, experience ourselves as a cosmos, as a beautiful order. Each new stage of life emerges from the annihilation of earlier stages. 'Antagonism' is far too weak a word to describe the contradictions that every living human must process internally. There is no self from which we could become estranged, for it was an unforeseeable chain of lethal estrangements that produced the self we supposedly are. But it is precisely because our self-identical nature is a fiction that we are always at a distance from ourselves. So we can know ourselves thanks to history, which is the present in us. An engagement with this history can uncover the future possibilities that are open to us. That is the historical nature of the human being.

The specific form of life that is disclosed for humans through their nature comes from the experience that, in their time, they feel that the anticipation of death moves them out of time. According to Pindar, they are 'the dream of a shadow' (Pythian Ode VIII, 99f.); the human being experiences itself as a creature that simultaneously is and is not. As we can only conceive of temporal things, the phrase 'outside of time' directs us into a void of thought. But the intra-temporal thought-form of negation is not applicable to this void of thought. On the contrary: we speak of this void because our behaviour, feelings and thinking convey to us that we float above it. We can, as Pindar's words show, experience it here and now. So we are thinking it in the wrong way if we think of it as a beyond. That is why we are breaking up the 'human condition' into pieces if we present death as the negation of life and life as the negation of death. Death is in the midst of life, life is in the midst of death. In our entire tradition, it is only in the New Testament that we find an experience of life in which the barrier between life and death is broken through. Jesus, it is said, overcame death.

In the same sense that Jesus of Nazareth was a human, we are only humans if, in affirming death, we do not flee from the depth of existence that is disclosed to us in the knowledge of time. But the human being to which I have just referred is no longer a specimen of a species. It is real, in its own experience too, precisely by being at home in the void of thought, by being *Homo absconditus*. That is why the historical nature of the human being must remain hidden from anthropology.

Part IV
Art and Myth (1972/1973)

Chapter 11
The Conflict Between Art and Society

The denizens of scientific-technological civilisation are not interested in the Christian faith or the truth of thought; they behave with indifference towards their antinomies.[1] They feel superior to the Christian faith because the autonomous human does not acknowledge anything to which they should bend the knee. The nature of truth is clear in the technological age: only that which functions is true. This does away with all the barriers that could limit the uppermost principle of this civilisation: whatever *can* be done should be done. The aforementioned indifference is starting to become transparent, however; it only thinly veils a conviction that carries this entire civilisation, namely the conviction that the human being in nature possesses a total monopoly of power. Doing away with faith and the truth of thought follows directly from this claim, for nothing that would challenge it can be tolerated. Whoever sells themselves to power is at the mercy of the play of powers. The sphere of powers is the realm of myth. Because our civilisation has done away with myth too, however, it no longer knows the magic spells with which the people of old were able not to control nature, but at least to build dams against its latent powers. Under the delusion of dominating all the powers of nature, of possessing them, human beings have squandered their freedom. All human life becomes a function of society, which in turn represents the totality of the human monopoly of power. But art also stands in irresolvable conflict with society. This statement, however, contradicts the mindset in which the consciousness of 'society' articulates itself.

The opinion that anything spiritual – religions, ideologies, art, even science – is a reflex of society has established itself with barely any resistance in the industrial age. Whatever cannot be derived immediately from the economic conditions, structures and power relations within the respective society is viewed as a mere epiphenomenon. The society in which we live cannot bear the existence of configurations that oppose

[1] This text was first published as: "Der Zwiespalt zwischen Kunst und Gesellschaft". In: Georg Picht, *Kunst und Mythos*. Stuttgart: Klett-Cotta 1986, pp. 18–35. The permission to include this text in this volume was granted by Klett-Cotta.

E. Rudolph and J. Picht (eds.), *Georg Picht: A Pioneer in Philosophy, Politics and the Arts*, Pioneers in Arts, Humanities, Science, Engineering, Practice 19,
https://doi.org/10.1007/978-3-030-31790-4_11

its compulsion to devour everything and adapt it to itself. It cannot bear that nature has its own rules, and therefore destroys ecological fabrics. What it can bear least of all is for constructs made by humans to defy the constraints of society. Any theory that reduces all spiritual creation to social relations is merely enforcing the diktat of that society's collective consciousness. Because science has indeed been reduced to a beadle of these collective tendencies in the twentieth century, however, the thesis that art is not a reflection of society, but rather stands in irresolvable contradiction to it, requires substantial explanation.

The marketplace of public opinion, especially in art education, is flooded with the waste products of such theory. That, to quote Marx, is 'beneath all critique'. Yet even at the highest level of critique, the assertion of a harmony between art and society is presented in a form that, as we have seen, is not immune to misinterpretation. Thus we encounter it in Theodor W. Adorno's *Aesthetic Theory*, a work in which more comes to light about the truth of art than in most artistic products from the same period:

> That society 'appears' in artworks, with polemical truth as well as ideologically, is conducive to historic-philosophical mystification. Speculation all too easily falls prey to the idea of a harmony between society and artworks that has been pre-established by the world spirit. But theory must not capitulate to that relationship. The process that transpires in artworks and is brought to a standstill in them is to be conceived as the same social process in which the artworks are embedded; according to Leibniz's formulation, they represent this process windowlessly. The elements of an artwork acquire their configuration as a whole in obedience to immanent laws that are related to those of the society external to it. Social forces of production, as well as relations of production, return in artworks as mere forms divested of their facticity because artistic labour is social labour; moreover, they are always the product of this labour. In artworks, the forces of production are not in themselves different from social productive forces except by their constitutive absenting from real society. Scarcely anything is done or produced in artworks that does not have its model, however latently, in social production. The binding force of artworks, beyond the jurisdiction of their immanence, originates in this affinity. If artworks are in fact absolute commodities, in that they are a social product that has rejected every semblance of existing for society, a semblance to which commodities otherwise urgently cling, then the defining condition of production, the commodity form, enters the artwork along with the social force of production and the antagonism between the two.[2]

If one views this text only superficially, and from the outside, Adorno seems to be fully and faithfully transferring the theory of the conditions of capitalist production, which Marx developed with great foresight in a long-past initial phase of the industrial age, to the realm of aesthetics. Then Adorno too would be a dogmatist who, like other exponents of Marxist aesthetics, imposed a predetermined theory immune to criticism on the art of our time, without asking how it might be possible to 'apply' a knowledge that comes from the Biedermeier period to the conditions produced by the technological-industrial revolution. But if one looks at the exact wording more closely and analyses the dialectical context, it becomes apparent that Adorno here only develops and presupposes the claim of a harmony between art and society in

---

[2] Theodor W. Adorno, *Aesthetic Theory*, trans. Robert Hullot-Kentor (London and New York: Continuum, 2002), p. 236.

order to show that this is *not* the truth demonstrated by his analyses. The objection to the doctrine of a 'pre-stabilised harmony between society and the work of art' is already built into the text. This doctrine is interpreted via the statement that the works 'windowlessly' (to use Leibniz's term) represent the social process in which they are embedded. 'Windowless'? Fifteen lines earlier, we read the following: 'The critical concept of society, which inheres in authentic artworks without needing to be added to them, is incompatible with what society must think of itself if it is to continue as it is; the ruling consciousness cannot free itself from its own ideology without endangering society's self-preservation.'[3] As a crystallisation of critical consciousness, works of art are anything but 'windowless'. Accordingly, the section ends with this realisation: 'In expression they reveal themselves as the wounds of society'.[4] Hence the doctrine that the work of art is a windowless representation of its agreement with the social process already contains within itself that contradiction which Adorno subsequently develops dialectically in his analysis of the *l'art pour l'art* principle. The same applies to the characterisation of artworks as 'absolute commodities' – a characterisation that Adorno introduces with a proviso. The work is 'absolute' through the autonomy it claims. A 'commodity', on the other hand, is nothing but reflected heteronomy. 'Absolute commodity' is a contradiction in terms, and the work of art, embodying this contradiction, is a 'wound'. This results in the aporia of art that Adorno develops in this section:

> If art cedes its autonomy, it delivers itself over to the machinations of the status quo; if art remains strictly for-itself, it nonetheless submits to integration as one harmless domain among others. The social totality appears in this aporia, swallowing whole whatever occurs.[5]

It is this aporia that must be revealed. That is why for Adorno, 'theory' is now only possible in the form of 'negative dialectics'. Although it presupposes the doctrine of the congruence between social and artistic production, it is not identical to it, but rather transforms the 'harmony' from which it proceeds into irresolvable dissonance. The identical element that, even in Adorno's thought, still binds the affirmative assertion of the agreement between social and artistic production to the dissonance subsequently developed from it, is the idea of society's 'totality', which remains inescapable even though it no longer produces, but rather 'swallows whole whatever occurs'. It is against this totality that the thesis of the conflict between art and society is directed. It proceeds from Adorno's aporia and will sublate it.

The doctrine that artistic production is always a reflection of social productivity and the conditions of production rests on an assumption whose dubiousness Hegel was already beginning to see through in his *Aesthetics*. It presupposes that the industrial production of commodities is 'production'. What does production mean?

In Plato and Aristotle, the 'concept' of production (*poiesis*) binds together natural philosophy, political philosophy and the philosophy of art. The later European tradition, dominated by Stoic forms of thought and a misinterpretation of Aristotle's

[3] Ibid., p. 236.

[4] Ibid., p. 237.

[5] Ibid.

theory of science, suppressed the problem of production. Through Kant's *Critique of Judgement* and Schelling's work, however, the power of the 'productive imagination' that carries all human thought and action was revived again shortly before the dawn of the machine age. Following on from the *Critique of Judgement* and Goethe, Schelling showed that we can only understand nature if it is transparent to us as production. Thus the idea of production became the axis of his entire system. It is constitutive of the mentality of industrial society, on the other hand, that, like everything else, it also attempts to usurp production; like a caustic solution, it dissolves the wealth of productive forms in the monotony of the socio-economic process. In the twentieth century, the word 'production' no longer evokes nature, literature or art, but rather factories. Theory adapted itself to this without resistance, albeit not innocently; but suppressing knowledge always has grave consequences. Both socialist and liberal economics only managed to 'produce' theories about the *relations* of production; no one asked about the nature of production itself. Until recently, one could delude oneself that such a suppression was acceptable, and that the order governing the *relations* of production also governs production itself in the desired fashion, which means in the interests of maximising production. But we know today that the exploitation of the earth's energy reserves and natural resources has limits, and that the destruction of nature caused by industrialisation threatens the existence of the human species. So we know that economic systems which measure their 'productivity' only by the quantity of commodities they churn out cannot last. If an economy is only capable of grasping the relation between the conditions of production and its quantifiable output, it has deprived itself of the regulatory mechanisms by which all growth in nature is kept under control. Its expansion *must* therefore violate ecological laws. We know, or could know, all of this. Because the conditions of production in nature have not been researched, however, and the entire domain of biological preconditions for economic productions therefore lies in obscurity, we know nothing about *what* must be produced, *in what forms* we may produce and *within what limits* the procedures for the technological-industrial manufacture of goods are possible, necessary or permissible. For all the constant talk of 'production', no one can say what this word is supposed to mean. This is so striking that it can only be explained by deep-rooted structural constraints within industrial society. The consciousness of such a society is clearly forced to suppress the question of the nature of production. In the industrial age, there is a conflict between society and production. The conflict between art and society is the symptom of a rupture in the foundations of the human condition that affects every sphere of human existence.

How can the conflict between industrial society and production be explained? The industrial manufacture of goods differs from pre-industrial economic forms primarily in the fact that human labour power has been replaced by machines on a previously unimaginable scale. The replacement of humans with machines has long ceased being simply a way to free humans from heavy physical exertion, for it now pervades all spheres of human activity. In the age of computers, it is even beginning to subject the activity of thinking to its own specific laws. It is in the nature of the machine that it cannot produce, but only reproduce. Mechanical fabrication is a reproduction of construction models. To the extent that the replacement of humans with machines is gaining ground, the basic human capacity for production is limited to those activities carried out in engineering offices and laboratories. But there too, what goes on is not production but reproduction; only things that increase the mass manufacture of mechanical reproduction are allowed to be constructed and 'invented'. As the whole of society – by no means only those who own the means of production – benefits from this profit maximisation, it has an overwhelming interest in stifling the realisation that mechanical reproduction destroys the true productive forces of human beings and robs them of the ability to shape their own habitat in harmony with its specific ecological conditions. Alienation is only secondarily the result of an unjust organisation of the *relations* of production; it stems primarily from the overgrowth of reproduction and the simultaneous suppression of human productive abilities. It thus stems from the technological-industrial methods of producing goods, regardless of the nature of the economic system that lives off machines and lends a society the specific character of 'industrial' society. It is only through reproduction that the produced goods become 'commodities' in Adorno's sense, for it allows consumer goods to be cast onto the market that are identical specimens of the same 'pattern', and whose monetary value can therefore be standardised. In this way, it is the industrial reproduction system that creates the preconditions for the quantification of all goods that is required for a rationalised exchange economy. But commodities are not mere objects; they preform their consumers' way of life. In the reproduction system of industrial society, life stories too are therefore transformed into interchangeable clichés of ready-made types. Now the lives of humans are no longer shaped by themselves. Their form is no longer that of production, but is imperceptibly made subject to the supremacy of the prevailing reproductive mechanisms. The rupture between reproduction and production is revealed as a rupture in life itself. What we call 'alienation' originates from this.

Let us ask once more: how was it possible for this deep rupture to be overlooked? A well-founded answer to this question would require an analysis of the history of industrial societies. But the sleight of hand whereby theory sought to jump over the chasm between production and reproduction is easy to identify: the two ways of producing goods are forced into the same overarching category. The name of this category, in Adorno's text and elsewhere, is 'process'. Every sequence in time is a process. If one claims that the 'process' as such has its own essence that obeys its own dialectical laws, one can effortlessly equate the production process, the reproduction process and the social process. Then one arrives at this statement by Adorno: 'The process that transpires in artworks and is brought to a standstill in them, is to be conceived as the same social process in which the artworks are embedded'.[6] We need not reflect here on how this conception of a historico-economico-social process was derived from Hegel's philosophy of history, and how it still manages to captivate minds as the ghost of a defunct metaphysics. Let us be naïve and ask: what is a process?

A process, as stated above, is a sequence in time. What is meant by a 'process' thus depends on how one understands time. An industrial society's understanding

[6] Ibid., p. 236.

of time is easily explained. Mechanical 'production' in all its forms is based on the application of energies that humans have turned to their purposes with the help of physics; every machine is applied physics. The devices of technological production must therefore be constructed in such a way that they obey the temporal understanding of the physics of inorganic nature. A technological-industrial civilisation only allows processes that can be described, calculated and controlled using physical methods. Industrialisation thus inevitably makes the whole of society subject to the physics that is used to construct its systems of production.

In physics, time is not represented dialectically but as a uniform linear parameter. Even without knowing the historical background of our modern understanding of time, it is easy to see how physicists came to represent time as a linear parameter, ignoring the differences between past, present and future. The faces of our clocks mirror the earth's rotation; we count the days on that basis. We measure the years according to the earth's movement around the sun, which means that our clocks and our calendars are an image of the movements within the planetary system. The fact that these movements are uniform and periodical means we can count days and years, as all periods are identical quantities. The simplest way to represent a series of numbers is to place them in linear order; hence the infinite straight line is the simplest way of mathematically projecting the periodicity of the earth's orbits. Because all humans rotate with the globe and all events in human history can be clearly assigned to a particular phase of its rotation, the movement of the planet provides an immutable system of reference to measure the time of sequences *on* the planet. Linear time is the time of chronology.

The fact that time can be measured in this way, however, need not mean that the nature of time and the primary experience of time are adequately expressed in the linear parameter of time. Humans *experience* time as life time. In life time, each section is a phase in the evolution of a life story. That is why every person, although their life can be measured using the numerical values derived from the rotation of the globe, has their own time, and within their life story, every day and every hour is unique and distinctive. What the dialectical philosophy of history considers 'process' is the autonomous sequence of an evolution, that is, the process of life that was understood as the act of self-production. Here the terms 'process' and 'production' were still being used in the nature-philosophical context in which they developed. But these terms can no longer be reconciled with the physical concept of time imposed on the whole of society by industrialisation, because physical time eliminates life time. Under the rule of physical time, the schema we use to measure sequences within inorganic nature is imposed on the temporal sequences of organic life, where the difference between past and future is constitutive. What had been a life process is forced to submit to the temporal mechanisms of inorganic 'processes'. Now life can no longer produce itself; the replacement of production by reproduction is therefore a direct result of the subjection of society to the diktat of physical time.

The thesis of the conflict between art and society has led us to the conflict between production and society. Reflecting on the predominant concept of time in this context opened up the even wider horizon of a conflict between nature and society. From the outside, society – like any other symbiosis between organisms – is something that appears within nature. It is not only each individual creature in a society that obeys the laws of nature; the forms of socialisation among these creatures are likewise subject to natural conditions, structures and functional contexts. Thus it seems a trivial matter that society is a piece of nature embedded in nature. As a result of the specific conditions we have considered, however, industrial society has developed a consciousness that perceives in nature only what it can 'appropriate' or, more precisely, what it can exploit for its own purposes. It reflects the entirety of nature into the murky medium of its own false consciousness. Thus a continuous autosuggestion cultivates the delusion that human existence in all its facets is exclusively determined by society. Desperation over the totality of society, which simply devours everything, is perhaps reminiscent of the joke about the drunkard who gropes his way helplessly around the advertising pillar and calls out, 'Now they've locked me up too!' But the spell that keeps the modern consciousness locked in the cage of its forms of reflection is no joke. For a society afflicted by delusions, this spell is as inescapable as all the other things it takes for realities. These realities too are self-produced; they are the mirror chamber of its delirium. Empiricism, which always rotates in this circle, can only confirm the autosuggestion that gave rise to the delusional system. If, on the other hand, one imagines that one can observe society from the outside as a phenomenon in nature, just as we study an anthill, one simply falls prey to a complementary delusion. For the observer is themselves a member of society, and therefore cannot claim the ability to catapult themselves out of the life context to which they belong in order to adopt an absolute perspective. The doctrine of the totality of society, then, has deeper roots than it might initially seem, especially when one has seen through its delusional character.

The intention of the critical theory developed by Horkheimer and Adorno was to do away with the constraints on consciousness that affected both society and the sciences of this society – not from the outside, but through system-immanent critique – and thus to render them visible in their dominance. Here the conflict between art and society soon came to light. For Adorno, aesthetics is at the centre of critical theory. He shows that subtle analysis of artistic forms can uncover what is actually happening to people and to the truth in the straitjacket of industrial society. This is only possible because the autonomy of art, which is now no longer freedom of the subject, but must rather be interpreted as the autonomous laws of art, proves so indestructible in the very fragility and refinement of its constructs that even the murderous 'physics' of industrial mass society cannot shatter them completely. The most delicate pen and ink drawing has the power to break the spell of the totality of society. If that is the case, we can no longer be content with this statement by Adorno: 'Social forces of production, as well as relations of production, return in artworks as mere forms divested of their facticity because artistic labour is social labour'.[7] This assertion requires an addition: we must ask what quality in the works themselves allows the situation Adorno describes to become transparent, letting in a light that exposes the spell of society as a delusion. A construct does not become a work of art simply by mirroring social conditions; it would be superfluous and meaningless as

[7] Ibid.

a mere reflection. What makes it a work of art is that it crystallises in a medium that lies outside of social constraints and can break through the hypnosis on which they are based. To be sure, the work is produced *in* society *for* society by a member of that society, but production appears in it with all its original force. It brings to light what otherwise remains hidden. It permits the manifestation of the scandal that there are emanations, forces, energies and sources of light in the world that are not part of an industrial society's reproduction system and unmask its immanent deception.

The old claim that all art is semblance must therefore be augmented with the statement that art cannot lie. Art is the most sensitive instrument of critique that we possess – assuming we can read it. Whoever fails to see that the concrete blocks of modern architecture are an expression of the brutality that constructs them is beyond help. Whoever cannot deduce from the linguistic form or the jargon of a prose text the mentality of those who produce such things deserves to be manipulated. Whoever can watch one of the Wagner productions popular today without noticing the lies demonstrates that this very lie meets their needs. Taste is a measuring instrument of mentality; the artistic form exposes the hidden thoughts, feelings and attitudes. The art of a society or a period reveals how it thinks. If that is the case, the prevailing mindset in the sociology of today must be inverted: then art is no longer a mere reflection of social consciousness, for this consciousness rather encounters, in art, criteria that it can neither adapt nor dominate, and which therefore express a power that is stronger than the consciousness itself. Then art cannot be interpreted in social categories, but society must certainly be judged according to aesthetic categories, even where it would never dream of such a thing. Art becomes a direct critique of society through its pure form, not its subject matter. It is not a critique of this or that society, but exposes the delusion and blindness that are constitutive of collective consciousness in every industrial society.

In the core of its phenomenality, which cannot be deformed by any change in style, art belongs to a sphere of human existence that, like nature, was overrun by industrialisation. In the industrial age, it therefore stands in an irresolvable contradiction to society – a contradiction that cannot be eliminated through society developing massive interests that seek to treat art like nature, that is, to appropriate, exploit and misuse it. Nor can this contradiction between art and society be done away with through the artists themselves – either because they can be bribed with success or because they have no idea – constantly making compromises with the dominant tendencies in society, which in turn damage their art. There have only been a few outstanding figures in the art and literature of the last hundred years who proved capable of asserting themselves in this murderous interrelationship with their integrity intact. Overall, the history of modern art confirms Adorno's statement that society swallows everything and knows how to use even those impulses that seek to end its dominance for its own purposes.[8] But does this not take us back to the doctrine of the totality of society?

[8] See ibid., pp. 236f.

Every stone, every blade of grass and every worm, every look at the clouds or the stars, teaches us the absurdity of claiming that our existence in its totality is nothing but a product of society. Only the thesis that our *consciousness* is a reflection of society merits theoretical discussion. In technological-industrial civilisation, the pressures to conform have become so overpowering in relation to the slim foundation of individual experience on which individuals could, in the best case, establish their independence, that the empirical findings seem to confirm the thesis that consciousness can now only be described as a collective phenomenon. But the assertion that social consciousness cannot transcend itself in its grasp of reality is not based on empirical evidence; rather, it is the result of a projection. Kant taught that transcendental consciousness cannot transcend its own boundaries. Neo-Kantianism took this doctrine out of its metaphysical context and turned it into an abstract epistemology; only thus was it possible to transfer the schematicisms of *transcendental* consciousness to *social* consciousness. Adorno presupposes all of this. He merges neo-Kantianism with the forms of reflection in Hegelian and Marxian dialectic. Admittedly, this sacrifices Hegel's absolute spirit and Marx's materialism, leaving only the infinite negativity of consciousness, namely pure subjectivity.

But now we must recall that Kant's theory of knowledge rests on a strict distinction between the transcendental and the empirical subject. The transcendental subject constitutes itself through a priori cognitions, that is, through cognitions that are necessary and universally valid, and hence retain their immutable truth at all times in all societies. The constitutive significance of the a priori cognitions thwarts any attempt to interpret consciousness as a reflection of empirical conditions and to equate social with transcendental consciousness. Such an attempt was enabled and encouraged by the fact that neo-Kantianism was based purely on the 'Analytic of Pure Concepts of the Understanding'. It was ignored that Kant himself supports the a priori content of our experiential knowledge metaphysically in his doctrine of transcendental ideas. But if one eliminates the metaphysics from Kant , his doctrine of the conditions of possibility of cognition changes into a theory of the functions of comprehension, where the difference between the transcendental and the empirical subject becomes irrelevant because the unconditional truth of cognition – however limited that cognition might be – can no longer be claimed. It is not so easy to get rid of metaphysics, however: if it is expelled from the centre of consciousness, it reappears on the periphery as the ghost of totality. The thesis of society's totality is therefore a projection of metaphysics driven into exile from the constitution of consciousness. At the same time, totality is the vehicle that in turn allows the forms of reflection from Hegel's logic of absolute spirit to be incorporated into this new theory of the subjectivity of consciousness. Totality proves to be the hollow mould of the absolute.

This whole projection of the formal structures of Kant's and Hegel's transcendental logic onto the consciousness of society falls apart, however, when one realises that transcendental logic and dialectic necessarily presuppose a distinction between the transcendental and the empirical subject. Through its constitution, consciousness is based on forms of cognition that transcend any possible society. But then one cannot claim that this same consciousness reproduces, in its totality, the relations of production in a particular society; the assertion contradicts itself. Either this theory, like every form of consciousness, is merely a reflection of certain social conditions – then one cannot claim it is true, for then it has already been made obsolete by the change in conditions. Or one does claim it is true – then it must transcend social conditions, and hence denies itself. That is why the transference of the schematicisms of transcendental logic to the empirical consciousness of society leads to a contradiction that inevitably thwarts it. Adorno himself asserts the inevitability of this failure in *Negative Dialectics*.

Today we cannot dogmatically cite Kant's explanation and limitation of the possibility of a priori cognitions; the great crisis of metaphysics has also affected the idea of the a priori itself. Transcendental logic and every possible form of dialectic, as we recognise today, necessarily imply metaphysics. But we can no longer rely on this metaphysics. On the contrary, we know that our consciousness is automatically constituted by cognitive forms and content that cannot be attributed to society. For consciousness – if only because it cannot be separated from language – is irrevocably collective memory. Though unaware of it, it preserves the memory of long-past phases of human history as well as very diverse forms of human coexistence. The irresolvable difference between its constitution as collective memory and its contingency on current social conditions are what make the coming-to-consciousness of human consciousness possible in the first place: consciousness is the experience of that difference. It is the experience of this contradiction that renders the adaptation collectively intended in this society impossible. One can see this in the concept of 'society'.

Like art, language contains potencies that bring to light what is untrue. Language too is not merely a reflection of society, for it contains our entire history. But it mirrors social conditions in all their facets. One can see this precisely in the use of the word 'society': it is treacherous with what thoughtless agility people use the word society; it has become a fetish. What does this word communicate? The concept of society developed in parallel with the emancipation of the bourgeoisie and industrialisation. In its essence, it is a term of bourgeois liberalism; it is steeped in the mentality of early industrial capitalism. One should not let oneself be deceived about this by the fact that the Left constantly utters this word. Socialism took its entire conceptual repertoire, its categories and its thought models from bourgeois theories, and is therefore at risk of denying itself with every statement it makes through the terms it uses. If the consciousness of an industrial society is trapped in the aforementioned 'context of delusion' (Adorno), then it is precisely the concept of 'society' that must represent its false consciousness of itself. Today one speaks of 'society' as if society were some hybrid of nature and fate; it is forbidden to look for something beyond 'society'. But society is neither God-given nor nature-given. All structures of human coexistence were conceived and produced by humans in the course of history. The society of the industrial age is one extreme on the spectrum of possible forms for organising human symbiosis, an extreme that only became possible with the construction of large-scale technological systems and a thoroughgoing change of natural living conditions. It is based on an artificial formation of the malleable material we call 'human' that is produced by the abuse of human freedom and contradicts, even destroys nature. Theories that are adapted to the mentality of an industrial society view all forms of organised human coexistence that have emerged over history as 'societies', and thus also project into the past the system of prejudices on which the self-image of this special modern form of society rests. They interpret the structures and inner dynamics of pre-industrial living conditions and political hierarchies based on abstract conceptions that were designed for the interpretation of functional processes within the technological-industrial reproduction system; at the same time, they reflect the malaise that characterises the mentality of intellectuals within these societies. These implications cling ineluctably to the concept of 'society' as it has been used since the beginning of the century. One would therefore be well-advised to limit the use of this term to the period that needed it for its own self-image. It applies to the Europe that has existed since the mid-eighteenth century; today it applies to all regions that have submitted to the mechanisms of technological-industrial economy exported from Europe. When we speak of the conflict between art and society, the word does not refer to some timeless entity, but must always be understood in this historical sense. The conflict between art and society has only erupted in its full force since the nineteenth century.

In other periods and cultures, art had an entirely different connection to the organisation of human communities. There are a large number of cultures in which artistic conceptions – usually conveyed through architecture – formed the basis of public and private life. Here art is constitutive for the order of life. If one examines the genesis of cities, cultural landscapes, overarching political orders and – in the Modern Age – states, one repeatedly encounters conceptions whose structures are no different from those of architectural designs. It is not a superficial metaphor when political theory constantly refers to the state as work of art – it *is* a work of art in its origins, and can be represented in the same aesthetic categories with which one also describes other works of art. If one goes back even further into the foundations of cultures and their political structures, one encounters sacred rites, mythical notions, divine images and cults. As if in X-ray images, beneath that surface we describe with the words 'politics' and 'society', this brings to light a succession of layers in which art and myth permeate each other. Art is in agreement with the political order when this order obeys the laws of production; it comes into conflict with the political order as soon as reproduction gains the upper hand. But even in the industrial system of reproduction, there is still a visible connection between artistic production and the organisation of communal life. It is not a metaphor when we describe the scientific-technological world as an artificial one. It encapsulates the heart of the matter, if this word also conveys that the construction of such a world violates the laws that govern true art. The contrast between what is artificial and what is art makes it clear that the conflict between art and society manifests itself in the sphere of art itself. If we ask philosophically what the nature of art is, then we are forced to break the taboos of social consciousness. We do not take society as the final authority, but rather try to uncover society's conditions of possibility once more. Perhaps this will also open up the prospect that society, having been created by humans, can one day be changed by them. We do not, at any rate, have the slightest chance of influencing the course of human history as long as we have no access to the laws of production and the origin of the productive faculty – that is, to the realm previously termed ‘myth’. The sphere of production and the sphere of myth lie deep in the substratum of our consciousness. Art is only the very end of the tip of the iceberg that protrudes from the surface of the water.

Chapter 12
Absolute Art and Politics

'The idea of the free totality of beauty'[1] was, as Hegel testifies, first formulated and asserted by Schiller.[2] Its content is different for Schiller and for Hegel, however; Schiller's *On the Aesthetic Education of Man* from 1795 is a political treatise.[3] Why does aesthetics here become politics? And what does the merging of these two spheres, so carefully separated in the bourgeois world, have to do with the concept of the absolute that was in the ascendant at the time? Schiller declares in the Second Letter that 'the most perfect of all works of art' is 'the building up of true political freedom.'[4] The 'political arena' in which, 'as [was] believed' in 1795, 'the high destiny of mankind [was] decided',[5] was France, the home of the revolution. This 'great action' dealt with the realisation of political freedom. 'A question which was formerly answered only by the blind right of the stronger is now, it appears, being brought before the tribunal of pure reason, and anyone who is capable of putting themselves in a central position, and raising their individuality to the level of a species, may regard themselves as an assessor at this court of reason'.[6]

The cause of the French Revolution concerns anyone who is capable of thinking as a citizen of the world. In 1795, however, the revolution had already failed; the hopes of an entire generation had been dashed. In the Fifth Letter, Schiller explains why the revolution was bound to fail:

---

[1] G. W. F. Hegel, *Aesthetics: Lectures on Fine* Art, trans. T. M. Knox (Oxford and New York: Oxford University Press, 1975), vol. 1, p. 62.

[2] This text was first published as: "Absolute Kunst und Politik". In: *Kunst und Mythos*, pp. 58–68. The permission to publish this text in this book was granted by Klett-Cotta.

[3] Friedrich Schiller, *On the Aesthetic Education of Man*, trans. Reginald Snell (Mineola, NY: Dover, 2004). The letters were written to Duke Christian von Holstein-Augustenburg in 1793 and published in 1795, in a heavily revised and extended form, in Schiller's journal *Die Horen*..

[4] Ibid., p. 25.

[5] Ibid.

[6] Ibid., p. 26.

E. Rudolph and J. Picht (eds.), *Georg Picht: A Pioneer in Philosophy, Politics and the Arts*, Pioneers in Arts, Humanities, Science, Engineering, Practice 19,
https://doi.org/10.1007/978-3-030-31790-4_12

> It is true that the deference to authority has declined, that its lawlessness is unmasked, and, although still armed with power, sneaks no dignity any more; men have awoken from their long lethargy and self-deception, and by an impressive majority they are demanding the restitution of their inalienable rights. Nor are they merely demanding them: on every side they are bestirring themselves to seize what has, in their opinion, been wrongfully withheld from them. The fabric of the natural State is tottering, its rotten foundations are yielding, and there seems to be a *physical* possibility of setting Law upon the throne, of honouring Man at last as an end in himself and making true freedom the basis of political association. Vain hope! The *moral* possibility is wanting, and the favourable moment finds an apathetic generation.[7]

One must call this text to mind word for word because, even today, we are still faced with the same problem that Schiller formulated in *On the Aesthetic Education of Man*.

Schiller's revolutionary utopia corresponds to that of Marx. Marx believed that a change in the relations of productions could create conditions under which 'true freedom' could become the 'basis of political connection'. We know today that this attempt too must be met with Schiller's words: 'Vain hope! The *moral* possibility is wanting, and the favourable moment finds an apathetic generation.' This makes it all the more fitting to ask today whether we should seek the solution to our current problems in that hidden area which Schiller sought to uncover with his treatise. With reference to this question, Schiller's analysis of the French Revolution merits further examination. He describes the drama that took place:

> Among the lower and more numerous classes we find crude, lawless impulses which have been unleashed by the loosening of the bonds of civil order, and are hastening with ungovernable fury to their brutal satisfaction. It may be that objective humanity had some cause of complaint concerning the State; subjective humanity must respect its institutions. Can we blame the State for disregarding the dignity of human nature so long as it was defending its very existence, for hastening to separate by the force of gravity, and to link together by the force of cohesion, where there could as yet be no thought of building up? The extinction of the State contains its vindication. Society uncontrolled, instead of hastening upwards into organic life, is relapsing into its original elements.[8]

To explain this disaster, Schiller, influenced by Rousseau, develops the theory of alienation that was subsequently taken further by Hegel and inherited by Marx. The collapse of modern civilisation is the consequence of a 'disruption, which Art and learning began in the inner man'.[9] 'It was culture itself that inflicted this wound upon humanity.'[10]

> The child of Nature, when he breaks loose, becomes a maniac, the disciple of Art an abandoned wretch. The intellectual enlightenment on which the refined ranks of society, not without justification, pride themselves, reveals on the whole an influence upon the disposition so little ennobling that it rather furnishes maxims to confirm depravity. We disown Nature in her rightful sphere only to experience her tyranny in the sphere of morality [...].[11]

---

[7] Ibid., pp. 34f.

[8] Ibid, p. 35.

[9] Ibid., p. 39.

[10] Ibid.

[11] Ibid., p. 36.

This explains the political condition of the modern world. The state becomes

> an ingenious piece of machinery, in which out of the botching together of a vast number of lifeless parts a collective mechanical life results. State and Church, law and customs, were now torn asunder; enjoyment was separated from labour, means from ends, effort from reward. Eternally chained to only one single little fragment of the whole, Man himself grew to be only a fragment; with the monotonous noise of the wheel he drives everlastingly in his ears, he never develops the harmony of his being, and instead of imprinting humanity upon his nature he becomes merely the imprint of his occupation, of his science. But even the meagre fragmentary association which still links the individual members to the whole, does not depend on forms which present themselves spontaneously (for how could such an artificial and clandestine piece of mechanism be entrusted to their freedom?), but is assigned to them with a scrupulous exactness by a formula in which their free intelligence is restricted. The lifeless letter takes the place of the living understanding, and a practised memory is a surer guide than genius and feeling.[12]

> [...] the State remains eternally alien to its citizens because nowhere does feeling discover it. Compelled to disburden itself of the diversity of its citizens by means of classification, and to receive humanity only at second hand, by representation, the governing section finally loses sight of it completely, confounding it with a mere patchwork of the intellect; and the governed cannot help receiving coldly the laws which are addressed so little towards themselves. Finally, weary of maintaining a bond which is so little alleviated for it by the State, positive society disintegrates (as has long since been the fate of the majority of European States) into a moral state of Nature, whose open force is only one *more* party, hated and eluded by those who make it necessary, and respected only by those who can dispense with it.[13]

Who can deny that this description of the modern state and its mechanisms of alienation anticipates a condition that would only come to fruition in the twentieth century? But the principle whereby Schiller deduces this disintegration of all developed customs and order is not that of capitalist economy, but rather the mechanistic mindset found in the bourgeois society that built up the modern state under the patronage of enlightened absolutism. The disruption of the forms of organic life, which, according to Schiller, also include historical constructs, is caused by the imposition on the community of a model derived from the physics of inorganic nature. The machine age begins, as Schiller testifies, with the construction of the state machine. The murderous processes of alienation triggered by mechanical production, that is, the transference of the same principle to the sphere of economy, are only the second step that presupposes the first, and then resulted in the second revolutionary movement, namely the socialist one. Both the bourgeois and socialist revolutions failed because they either failed to recognise the principle of disruption clearly named by Schiller, namely the mechanistic mindset, or, if they did recognise it, were not willing to deal with it. The bourgeois revolution adopted the mechanistic state apparatus of princely absolutism and turned it to its own purposes. The socialist revolution adopted both this state apparatus and the industrial complexes built up by capitalism. In both cases, the revolution soon degenerated into a coup that brought a new ruling class

[12] Ibid., p. 40.

[13] Ibid., pp. 41f.

to power and cheated society of the freedom it had been promised. Schiller already saw this in 1795 with merciless clarity:

> The old principles will remain, but they will wear the dress of the century, and philosophy will lend its name to an oppression which was formerly authorised by the Church. Terrified of the freedom which always declares its hostility to their first attempts, men will in one place throw themselves into the arms of a comfortable servitude, and in another, driven to despair by a pedantic tutelage, they will break out into the wild libertinism of the natural State. Usurpation will plead the weakness of human nature, insurrection its dignity, until at length the great sovereign of all human affairs, blind Force, steps in to decide the sham conflict of principles like a common fistfight.'[14]

A few years after these lines were written, Napoleon Bonaparte seized power.
One cannot fight this dynamic with reason:

> Reason has accomplished all she can, in discovering and expounding Law; it is the task of courageous will and lively feeling to execute it. If Truth is to gain the victory in the struggle with Force, she must first become herself a *force*, and find some *impulse* to champion her in the realm of phenomena; for impulses are the only motive forces in the sensible world.[15]

The rational Enlightenment of the eighteenth century closed its eyes to this circumstance, as did its successors in the twentieth century. That is why the European Enlightenment repeatedly came to a halt.

> It is, therefore, not enough to say that all intellectual enlightenment deserves our respect only insofar as it reacts upon the character; to a certain extent it proceeds from the character, since the way to the head must lie through the heart. Training of the sensibility is then the more pressing need of our age, not merely because it will be a means of making the improved understanding effective for living, but for the very reason that it awakens this improvement.[16]

Hence true enlightenment which alone, 'under the influence of a barbarous constitution',[17] could open up the sources of a renewal is not to be found in science, but rather in the development of sensibility. The instrument that achieves this is 'the Fine Arts'.[18] Art alone can reveal to us the models for the harmony of reason and inclination, intellect and sensuality, sentiments and drives that is the prerequisite for freedom in both the moral and the political sense. The pictures in art present us with models for the reconciliation of the individual with the general, which is what distinguishes a liberal state order from a mere state of nature. By awakening the drive to beauty, understanding such models is what opens the human spirit up to reason in the first place. This gives rise to the starting thesis of Schiller's letters, namely 'that we must indeed, if we are to solve that political problem in practice, follow the path of aesthetics, since it is through Beauty that we arrive at Freedom.'[19] Thus politics turns into aesthetics and aesthetics into politics. They are held together by

---

[14] Ibid., p. 47.

[15] Ibid., p. 48.

[16] Ibid., p. 50.

[17] Ibid.

[18] Ibid., p. 51.

[19] Ibid., p. 27.

the idea – directed against Kant's moral philosophy – that freedom only comes about when intellect and sensuality, reason and drives are reconciled in a higher harmony, and where the difference between theory and praxis is dissolved in the unity that constitutes the originary 'humanness' of humans. According to Kant, awareness of freedom is the only essential content of human reason. If the realm of beauty opens up above the freedom that was encapsulated in the categorical imperative, then beauty is the absolute as such. Then art is the representation of the absolute, and art itself becomes absolute art.

The idea of absolute art is, as it transpires, falsified at its very roots if it is made to serve the purpose of severing art from politics, from the struggle to realise freedom in history, and to remove it to an insubstantial sphere beyond the great concerns of human existence. If 'it is through Beauty that we arrive at Freedom' – at *political* freedom – then the slogan *l'art pour l'art* denies the absolute content of art. Rather, by having to realise the idea of absolute art, the artist receives a political mission. Their task is to wage the battle against the political disruption of our civilisation, which places them in irreconcilable opposition to the state and society.

> No doubt the artist is the child of his time; but woe to him if he is also its disciple, or even its favourite. Let some beneficent deity snatch the infant betimes from his mother's breast, let it nourish him with the milk of a better age and suffer him to grow up to full maturity beneath the distant skies of Greece. Then when he has become a man, let him return to his century as an alien figure; but not in order to gladden it by his appearance, rather, terrible like Agamemnon's son, to cleanse it. He will indeed take his subject matter from the present age, but his form he will borrow from a nobler time – nay, from beyond all time, from the absolute unchangeable unity of his being. Here, from the pure aether of his daemonic nature, flows forth the well-spring of Beauty, untainted by the corruption of the generations and ages which wallow in the dark eddies below it.[20]

Already in the 1788 poem 'The Artists', Schiller had praised beauty as the manifestation of the absolute:

> For we shall recognise as truth at last,
>
> what here as beauty only we have viewed.[21]

But what is missing here is the political dimension of art: the merging of the idea of beauty with the idea of freedom. The poem was written before the French Revolution. With the letters *On the Aesthetic Education of Man*, the artist entered the political arena of his time. Only now did it become clear that the idea of absolute art was not only a projection of poetic enthusiasm, but that it also revealed the true relations between the state and society with a clarity that elevated Schiller above all the political thinkers of his era. His sacrifice for this purpose is an idea that had informed European poetics since Horace and was still fundamental to the poem 'The Artists': the idea that art exists to please humans and reconcile them to their fate. Now the artist is 'an unfamiliar figure' in his own century; he comes 'not to delight

---

[20] Ibid., pp. 51f.

[21] Friedrich Schiller, *The Poems of Schiller*, trans. Edgar A. Bowring (Charleston: BiblioBazaar, 2009), p. 78.

it with his appearance, but as terrible as Agamemnon's son, to purify it.' One could consider it a trial by fire for the authenticity of the idea of absolute idea of art, perhaps, that Schiller was willing to sacrifice to it all that had served as a justification for art until the late eighteenth century: decoration, lightness, seduction, glamour and all the sweet gifts for which the muses were praised. If art is absolute, it must disdain the semblance of reconciliation. Its task is that of Orestes: purification and vengeance. This thought too expresses a pitiless recognition of historical reality: art as a lovely embellishment of existence perished in the French Revolution, never to be revived. In the nineteenth and twentieth centuries, the theme of all great art was the exposure of society's lie; its formal law was dissonance, not harmony. The great artists and the great poets reject the notion that they are favourites of their time. Such figures as Baudelaire or Stefan George perfectly fit Schiller's description of the artist, thus proving his startling modernity.

We have let Schiller speak for himself, as no one would otherwise believe in which context the idea of absolute art arose. Schiller's definition of art emerges from the assertion at the start of the letters that 'the most perfect of all works of art' is 'the building up of true political freedom.'[22] It breaks out of the sphere of the 'fine arts'. In everything Schiller says about artists, he has in mind the 'pedagogical and political artist, who has Man at the same time as his material and as his theme.'[23] The artist's goal is the transformation of state and society that thwarted the French Revolution:

> If, therefore, the principles I have laid down are correct, and experience confirms my description of the present time, we must continue to regard every attempt at reform as inopportune, and every hope based upon it as chimerical, until the division of the inner Man has been done away with, and his nature has developed with sufficient completeness to be itself the artificer, and to guarantee reality to the political creation of Reason.[24]

He finds the guarantee of the possible reality of the political art that he demands in two fundamental laws of the sensual-reasonable nature of humans:

> The first insists upon absolute *reality*: he is to turn everything that is mere form into world, and realise all his potentialities; the second insists upon absolute *formality*: he is to eradicate in himself everything that is merely world, and produce harmony in all its mutations; in other words, he is to turn outward everything internal, and give form to everything external.[25]

From this, we can deduce what the concept of the absolute means for Schiller: 'Both tasks, considered in their supreme fulfilment, lead back to the conception of divinity from which I started.'[26]

If one compares this idea of art to Hegel's doctrine of beauty as the sensuous semblance of the idea, one sees at first glance what restrictions the later Hegel

---

[22] Schiller, *On the Aesthetic Education of Man*, p. 25.

[23] Ibid., p. 33.

[24] Ibid., p. 46.

[25] Ibid., pp. 63f.

[26] Ibid., p. 64.

imposed on Schiller's thoughts, despite invoking them. Art is presented as a manifestation of the absolute, yet simultaneously limited strictly to the system of the 'fine arts' once more. Thus isolated, it lacks the power to influence the sphere of the 'objective spirit', that is, of law, morality and the morals realised in the family, bourgeois society and the state. On the contrary: Hegel's theory of the end of art relies substantially on the insight that the historical development of the objective spirit in society and the state leave no space for the appearance of the absolute in the form of art. The same applies to religion, even to philosophy, whose last word is the tragic realisation from the preface to the *Philosophy of Right*: 'When philosophy paints its gray on gray, then has a form of life grown old, and with grey on grey it cannot be rejuvenated, but only known; the Owl of Minerva only takes flight with twilight closing in.'[27] These words were written in 1820. They express the profound resignation that took hold of intellectual Europe after the Carlsbad Decrees. They convey nothing less than the fact that the knowledge of the absolute cannot change historical conditions. Not only art, but also the absolute spirit as such became powerless, capitulating to the inexorable autonomy of the mechanisms that the state, society and business obey in the industrial age. But if spirit can no longer 'rejuvenate', no longer produce life, the talk of absolute spirit loses its truth. Hegel's theory of the end of art concealed his realisation, never stated openly, that metaphysics itself had reached its end. In the sphere of thought, all this follows with merciless necessity from the separation of art and politics, the separation of beauty and political freedom. It was self-deception on the later Hegel's part to think that he could lay out the absolute nature of art while restricting it to one particular area of public life; this reduces absolute art to mere 'art', whose consignment to a museal sphere doomed to historical ineffectuality can be supported and justified not through the absolute concept, but only through established forms of intellectual thought. But this isolation and exclusion of art corresponds exactly to the actual course of history, which ultimately forced art into the absurd situation of seeking to justify its absolute *aspiration* by the same separation from politics that had robbed it of its absolute *nature*. Schiller was more consistent in this regard: because art makes the absolute visible, he places it in truly 'absolute', that is to say, irreconcilable opposition to the historical world of the Modern Age. Now the great power of reconciliation can only gaze upon this historical world with the implacable face of Orestes. Its purpose is not to please but to purify and avenge. It appears as absolute negation. Whenever great art has achieved a breakthrough in the nineteenth and twentieth centuries, in opposition to Hegel's doctrine, it has had this negative character. It brings to light the antinomies and dissonances that are concealed by the state and by society. It does not beautify existence but unmasks it, and by this very act safeguards the unfading value of beauty. In the twentieth century, the power of the absolute manifests itself whenever art flies in the face of everything that society expects of it. Then it is a political power, as totalitarian systems have always known; only the liberals were so clueless that they did not perceive the political explosivity of elemental art. We therefore stand by the paradoxical conclusion

[27] G. W. F. Hegel, *The Philosophy of Right*, trans. T. M. Knox (Oxford and New York: Oxford University Press, 1967), p. 13.

that whenever art achieves a representation of the absolute, and thus allows it to irrupt into society, this absolute automatically makes it a political power. Whenever it turns its back on politics, it is simultaneously denied access to the absolute; it becomes powerless and artificial. It degenerates into mere decoration. Art is not political by virtue of presenting political topics, reproducing political opinions or allowing itself to be abused for political propaganda; it is political by virtue of being absolute. What makes it political is not its thematic content but its form.

Chapter 13
The Crisis of European Art and the Emergence of Esthetics

Around the middle of the eighteenth century, Europe – in parallel with the circumstances preceding the French Revolution – saw a rejection of the established European art traditions whose radicality and depth we are no longer conscious of because the educated bourgeoisie of the nineteenth century had an interest – despite being unaware of it – in concealing it.[1] It can be explained historically by the fact that the predominance of those powers that had used the arts as one of their most important means of representation in the Middle Ages and the early Modern age began to falter: the church and the nobility. To be sure, art forms had already developed in the age of humanism that served to represent the new self-confidence of the bourgeoisie and were connected to the bourgeois forms of Protestantism; they enjoyed exclusive power in the Netherlands as early as the seventeenth century. Wherever the Counter-Reformation triumphed, however, it repressed not only the new form of faith and its bourgeois attitudes, but also its expression in art. Yet art cannot lie. The great decorative style of the late Renaissance and the Baroque transformed imperceptibly into a deliberate illusionism, the visual representation of clerical deception. From the mid-eighteenth century onwards, this led to the destruction of all art forms whose expressive content was, in the public consciousness, inseparable from ecclesiastical hierarchy and displays of feudal or princely splendour. People discovered that art belongs to humanity, and that it becomes untrue when it degenerates into the privilege of a ruling class.

This shift manifested itself on the surface of the aesthetic consciousness in the fact that the sensibility of taste developed in Rococo art turned inwards and discovered the infinite nuances in the play of feelings. The immediate language of the heart was experienced as the language of nature, listened to and given poetic expression. But this silent and subtle transformation of artistic memory retention brought about

[1] This text was first published as: "Die Krise der europäischen Kunst und die Entstehung der Ästhetik". In: *Kunst und Mythos*, pp. 70–77. The permission to include this text here was granted by Klett-Cotta.

E. Rudolph and J. Picht (eds.), *Georg Picht: A Pioneer in Philosophy, Politics and the Arts*, Pioneers in Arts, Humanities, Science, Engineering, Practice 19,
https://doi.org/10.1007/978-3-030-31790-4_13

a revolution: it led to an abandonment of the poetic rhetoric whose artistic forms had defined the formal syntax not only of European literature, but also European music, from antiquity until the end of the eighteenth century, and whose effects on the formal design of the visual arts cannot be overstated. It was not only in the twentieth, but already in the late eighteenth century that the basic formal structures of European art began to waver. What had previously been art now seemed artificial. The representative style that, under Roman influence, had determined the entire spectrum of stylistic possibilities in the European arts seemed like a falsification of nature. Through its contrast, the sphere of inwardness and subjectivity opened up by bourgeois religiosity made the world of traditional art forms appear external, and thus untrue. The new world perception that stirred here asserted itself in seemingly very disparate areas. Sentimentalism, *Sturm und Drang*, pietism and the ideas of Rousseau all belong within a single context that is inseparable from bourgeois emancipation movement and resulted in the European consciousness turning away from everything that European art had represented from antiquity to the Baroque.

Pietist religiosity and new sentimentalism, the discovery of individuality and the language of the heart, are such delicate phenomena that one can barely comprehend now how it was possible for this emotionality to seize and reshape the whole of Europe. We have lost the sensitivity to the political effects that can result from a change in the forms of artistic representation. But even today, the shift that took place can be understood. In the Reformation, an elemental experience of immediate human proximity to God had asserted itself and created the forms of faith appropriate for it. Religion had been transferred to the conscience; yet the conscience of Protestant religiosity became the breeding ground for subjectivity. A human immersed in their own subjectivity no longer discovers God within their feelings, but rather themselves; they seek the same immediacy in their relationship with themselves that had been experienced so passionately in the Reformation as the immediacy of their relationship with God. This immediacy was now demanded for the relationship between humans too. The whole giant edifice of class structures and hierarchies that had told people their place and defined their self-identity in the Middle Ages and early Modern Age now appeared as a system of untruth and distortion; it began to crumble. But the new world feeling did not stop there: in the immediacy of subjectivity to itself it also discovered its immediacy to nature. That is why the opposition between nature and civilisation erupts in Rousseau. The whole world of culture and education was now presented as a product of alienation, as a falsification of human nature, as nothing but a great lie. And the art that had existed until then had belonged to this world of lies in so far as it did not, like the folk song, allow the original voice of nature to sound. On every page of Rousseau's work, one can sense how the over-refined sensitivity of a taste individualised by eighteenth-century art, by turning towards the immediacy of the heart, finds itself in a field of elemental forces that will tear everything down if they erupt. These forces can no longer be individualised; they exceed human existence. They belong to the universe. But as soon as art liberates itself from the individualism of the eighteenth century and no longer thematises the sensibilities of its own heart, but rather nature as such, humanity and the spirit of the universe, it can then define itself only as representation of the absolute, as absolute art. In the medium of new immediacy, as an absolute, it pushes towards immediate presence. The immediate presence of the invisible cannot be represented in the media of the visual arts, namely shape and colour. Hence the idea of absolute art can only be realised in literature and music. For the other arts, the 'end of art' invoked by Hegel had already taken place before the French Revolution.

There is a text in which we can see that the revolution taking place here in art actually extends much further. In Jean Paul's first novel, *The Invisible Lodge*, published in 1792, a passage in Sector 27 under the heading 'Gustavus's Letter' features a description of the Romantic hero, Gustavus, going to the park with his drawing board: 'I seated myself with my drawing pen upon an artificial rock near the lake and prepared to draw the scene [...] but when all Nature in full height sat to me and dazzled me with its rays that reach from sun, to sun, then did I adore what I would have copied, and sank at the feet of the Goddess and God...'[2] In the poet's experience, this is the artistic end of painting: 'I rose with lamed hand and surrendered myself to the sea which bore me up.'[3] This is followed by a great pantheistic hymn to the infinite unity of life in nature: 'Today I became linked to the whole creation and gave all beings my heart...'[4] But this is not yet the highest level. The hero turns eastwards; he turns back towards the 'set sun' and thinks of how he had once taken it for God, and is struck hard by the realisation that he has thought of God so rarely. Now he has left the goddess Nature behind. He rises alone above the creeping things and feels

> what man feels, but only he upon earth – when I could kneel in solitude with human eyes before the great mirror stretching out into nothingness, at which the insect dashes his feelers, before the mirror out of which flames the infinite giant, the sun... No! in earthly colours and on a canvas of animal skins and on all that lies before me, is merely the *image* of the arch-genius; but in man is not His image, nay, it is Himself...[5]

The picture painted in 'earthly colours' is true nature, that same nature glorified as a goddess in the pantheistic hymn. The 'canvas of animal skins' is a classic Jean Paulian metaphor for the surface on which true nature represents itself as the image that, as the hero now recognises, it is. That is why he says 'on all that lies before me, is merely the *image* of the arch-genius'. This corresponds to the underlying idea of the novel, which emerged from a vision of death: that this world, which has fallen prey to impermanence and death, is like a cloud concealing the true world, to which we will cross over in death.

Here the fate of art is prophetically anticipated. One of the basic forms of art, the image, is shattered in two successive stages. First Gustavus's hand is paralysed by the divinity of nature, which it had meant to represent in an image. He surrenders to the rising sea, which lifts him up, and attains unity with the whole of creation. In a second stage, the creation too is revealed as a mere '*image* of the arch-genius; but in man is not His image, nay, it is Himself...' Now nature too has become a mere image, and thus untrue. For only the immediate is true, and no longer the image, which merely

---

[2] Jean Paul, *The Invisible Lodge*, trans. Charles T. Brooks (Frankfurt: Outlook, 2018 [1883]), p. 173.

[3] Ibid.

[4] Ibid., p. 174.

[5] Ibid., p. 175.

represents it. But only human subjectivity is immediate in itself; in its immediacy it experiences God. What becomes of art when all images are shattered? Where does it now take the poet? 'Half of the sun still blazed above the rim of the earth, which cut it asunder; but I saw it no more through my dissolving eye – smothered, sunk, extinguished as I was, in the sweeping, flaming, rushing, shoreless sea around me...'[6] This is the experience of the absolute whose representation in visual art is no longer possible, for this experience denies the image as such. It carries the new literature and music as its infinite content. Towards the end of the nineteenth century, Mallarmé experienced the imagelessness of the absolute in comparable fashion. Then, in the radius of his poetry's influence, 'non-representative' art was born.

Unlike Mallarmé, Jean Paul was not brought to the edge of silence by his experience of the absolute. In his case, the dissolution of the visible world and the earth, with all that it carries, into mere images that rise and disappear again in time, transformed all that exists into metaphors and made every phenomenon comparable to every other, allowing those cascades and fireworks of unfamiliar, surprising, confusing and illuminating comparisons, symbols and analogies that are unparalleled in our literature to rise up. The artistic statement no longer lies in any determinable thematic content, but rather in the movement of a lyrical-musical stream that, with its infinite melody and its immense escalations, already seems to anticipate the music of Wagner. The truth of this melody, floating above infinity, is confirmed by its contrast – always shrilly underlined – with the base, contemptible or comical nature of the social and political circumstances from which this voice rises. Despite the literary models that informed Jean Paul's work, this is an art form that broke away from the hitherto binding rules of European poetics. He should therefore be assigned the first place in the genealogy of modern art and literature.

The dissolution, shattering and sinking of images and of everything that is mere appearance was dialectically elaborated in Hegel's philosophy as absolute negativity. The constitution of infinite being-with-itself, in which the spirit assures itself of its identity with the 'arch-genius' after the undoing of all human alienation, is defined in the philosophy of German Idealism as freedom. Freedom is the positive content of the absolute, and manifests itself through its absolute negativity. Freedom is 'absolute' in the sense that, as Jean Paul's text also shows us, it has cast off the shackles of phenomenal nature and is now with itself. Only then does the nature of beauty reveal itself to it. But this beauty now lies below it. Hegel writes in the introduction to his *Aesthetics*:

> But what is *higher* about the spirit and its artistic beauty is not something merely relative in comparison with nature. On the contrary, spirit is alone the *true*, comprehending everything in itself, so that everything beautiful is truly beautiful only as sharing in this higher sphere and generated by it. In this sense the beauty of nature appears only as a reflection of the beauty that belongs to spirit, as an imperfect incomplete mode [of beauty], a mode which in its *substance* is contained in the spirit itself.[7]

---

[6] Ibid.

[7] G. W. F. Hegel, *Aesthetics: Lectures on Fine Art*, vol. 1, trans. T. M. Knox (Oxford and New York: Oxford University Press, 1975), p. 2.

This follows from the definition of the absolute as freedom. It follows from the idea of absolute subjectivity, whose origin in God's immediacy for the faithful conscience in Protestantism it was therefore necessary to recall.

If humans only become certain of the absolute truth and their own freedom in this truth through the sensation of themselves, through 'feeling', then beauty must shift from the externality of its appearance to the feeling of beauty itself. The recognition of beauty is then no longer the discovery of what, outside of ourselves, we marvel at as something perfect; the external is only a reflection of the internal, whose hidden depth is unlocked and revealed to it by a feeling for beauty. In the feeling for beauty we feel as we are with ourselves, released from all external ties and hence free and 'absolute'. In gazing at the work of art, the human discovers, as if in the mirror, the ability – hidden in the inwardness of their feelings – to be put in the floating state of this harmony and its perfection. Then the space of beauty is no longer the world, but rather the interior of our mind as disclosed through our feelings. That is why, since the publication of Baumgarten's *Aesthetica* (1750–1758), the doctrine of the recognition of beauty changed into a doctrine of sensation. When we refer to the philosophy of art by the name 'aesthetics' (which Schelling and Hegel rejected), we are operating within this tradition. We are then seeking insight into the nature of art not in the works, but in the subjects that sense their beauty. In their work, the artist does not represent something real that lies outside of them; rather, they use the medium of their individuality to *express* human subjectivity as such. A conception of art that wishes to rediscover and understand its own subjectivity must endeavour to decode art per se as an expression of the subjectivity mirrored in it. We have seen how this view of art established itself in conjunction with the social and political changes and revolutions in the seventeenth and eighteenth centuries until, around the dawn of the nineteenth, it attained full consciousness of its own nature in the idea of absolute art and its absolute subjectivity. But this caused art itself to change, for artists are also affected by this process. The truth of art now no longer lies in the work, or in what the work seeks to represent; the truth lies rather in the subjective capacity, which is only mirrored in the work. Art as such, the absolute capacity, is now recognised as the thing that is meant to represent itself. The process of feeding back into the feelings of the subject, of re-flection, can no longer be separated from the work of art; it is incorporated into the character of art. This tendency reached its culmination in modern art: now poetry itself became the content of the poem, painting became the content of paintings and composing became the content of composition. Now the works themselves have also become 'aesthetic'. They present themselves as aesthetic objects. But all this, as we have seen, is a consequence of the idea of absolute art as the expression of absolute subjectivity.

The human being whose aesthetic education, according to Schiller, is meant to build the edifice of political freedom as the highest work of art, is identical to the human being who, when gazing at art, feels the inner harmony of the free play of their senses and reflects upon this feeling in aesthetics. They raise themselves, in the words of Jean Paul, 'above the creeping things [...] and feel what man feels, but only he upon earth'.[8] In aesthetics, this feeling enters the thinking self-consciousness; now the human recognises in themselves not merely the image of the arch-genius, but rather the arch-genius itself. With this, they have made the transition to absolute freedom. But the abandonment of the image has taken them beyond art: the completion of aesthetics is the end of art. The human being experiences unity with God in the absolute freedom of infinite subjectivity as *spirit*. In retrospect, they grasp the history of the human race as a progressive realisation of the idea of freedom; they grasp it as a history of spirit. This new concept of history mediates between the political and aesthetic aspects of absolute art. It gives voice to that new experience of the world which found its highest manifestation in the idea of absolute art.

[8] Jean Paul, *The Invisible Lodge*, p. 175 (translation modified).

Chapter 14
Myth and Affect

14.1 Phobos – God of Terror

If we go further back in the history of the theory of affects, we discover that the description of these phenomena has already taken us to the middle of the field of myth.[1] The majority of those phenomena later described in philosophy as 'affects' appear in a very different form in Greek poetry and literature before Plato. They appear as gods or divine powers. Defining the relationship of art to the affects thus leads directly to the question of the relationship between art and myth. What we are supposed to think of when we hear the word 'myth' is unknown for now. What people call 'affects', however, is something we suffer daily. Here we find ourselves in an area where we know from experience, or can yet learn, which phenomena were given divine names by the Greeks. We can ascertain how mythical thought and mythical experience of the world interpret phenomena that we ourselves perceive, and we can critically examine which of the two interpretations are more suited to the phenomena. It is therefore not a waste of time for me to illustrate what I have said with a simple example.

In Book IV of the *Iliad*, Homer describes how the Greek and Trojan armies advance on each other. The Trojans are driven on by Ares, the Greeks by Athena. In addition to these two gods, however, another two are also at work without any affiliation to either army, instead appearing between them: Deimos, Phobos and Eris.[2]

---

[1] This text was first published as: "Mythos und Affekt". In: *Kunst und Mythos*, pp. 441–457. The permission to republish this text here was granted by Klett-Cotta.

[2] 'Thus the clamour, from the immense Trojan muster,
Not one voice, one language,
But a cacophony of tongues from different lands.
Behind them, Ares, as behind the Greeks
The goddess Athena with sea-grey eyes,
And on both sides Terror [*deimos*] and Panic [*phobos*]
And Ares' murderous sister, Eris,

E. Rudolph and J. Picht (eds.), *Georg Picht: A Pioneer in Philosophy, Politics and the Arts*, Pioneers in Arts, Humanities, Science, Engineering, Practice 19,
https://doi.org/10.1007/978-3-030-31790-4_14

What is peculiar about the names of these gods is that they also appear in language as common words, ones we still use today: Deimos is fear, Phobos is the terror that can put an army to flight and Eris is discord. We will concentrate here on Phobos, because he also has a fixed place in the later theory of affects.

First of all, we can note that there is no doubt that Phobos is a god, in every possible sense of the word. There is clear evidence that there were cults of Phobos, such as a ritual inscription from the fifth century BCE that was carved by the Selinuntines after a victory. Part of it reads: 'Through Zeus we triumphed and through Phobos and through Heracles and through Apollo'. At the beginning of *Seven Against Thebes* by Aeschylus, the messenger reports how the seven commanders swore to conquer the city or die. In an undoubtedly ancient sacrificial ceremony, they slaughter a bull and collect its blood in a black shield. The commanders dip their hands in the blood and call on Ares, Enyo and the 'bloodthirsty god of Terror' to witness their oath.[3] When Plutarch reports that Alexander the Great sacrificed a bull to Phobos with mysterious ceremonies the night before the Battle of Gaugamela, when all the Macedonians were sleeping, accompanied by a seer, this was the same cultic act.[4] Phobos was especially revered in Sparta as a god of war. The purpose of this worship is clearly evident: Phobos is the one who puts the enemy army in a state of panic and terror. What is meant when the same word used by Greeks in the understandable and familiar meaning of 'terror' appears as the name of a god to whom one sacrifices, and who is worshipped in cultic rites? If you consider the corresponding verb *phobeō*, you will find that its active meaning is 'fill with terror', 'put to flight in panic'. In the passive and middle voices, it means 'be seized by terror', 'flee', 'be horrored', 'be afraid'. We can see from this usage how a single phenomenon can be viewed from two sides, just as the god Phobos appears between the two armies. The threat posed by an army cannot be separated from the terror and panic it causes the other army, for these effects are reciprocal. Both armies are exposed to the twofold effect of threat and terror. Homer thus describes the situation entirely accurately: the two armies close in on each other, something arises between them that we would today call suspense; it grows with each step, and is sensed by both parties as an ambivalent power. In the zone of this almost unbearable suspense, each party at once threatens the other and is terrified by it. No one can foresee in which direction the suspense will explode and which party, seized by panic and terror, will have to flee. One thing is clear, however: the suspense that builds up between the two armies is something completely different from an affect in the heart of each individual warrior. Anyone who has witnessed a crowd being seized by panic knows that such panic is experienced by every individual involved as an outside force that comes over them, a force they cannot resist and by which they feel overwhelmed, and which forces them to act against their own feelings,

---

Small when her crest first appears,
But so ravenous and relentless in her ways
That she soon thrusts her head into the sky
Even while she keeps her feet on the earth.' (434–443) Homer, *Iliad*, trans. Stanley Lombardo (Indianapolis: Hackett, 1997), p. 78.

[3] l. 44–45.

[4] Plutarch, *Life of Alexander*, 31,9.

their own insight and their own character. Once the panic is over, they say, 'I have no idea what happened to me.' This is then termed a mass psychosis, as if the crowd had a collective soul that obeys the same mechanisms as the soul of the individual. But if one enquires more closely what the word 'psychosis' is actually intended to mean here, who is the subject of this psychosis and where the soul that suffers this psychosis is located, no one knows the answer. In the example with the two approaching armies, moreover, one is dealing not with one but with two opposing crowds. If one removes one of them, the collective soul disappears along with the psychosis. In reality, the situation is rather exactly as Homer describes it: Phobos is a power that appears in the middle between the two armies when they close in on each other. Because each warrior in each army experiences that this power is stronger than he is, the Greeks called it 'god' – for whatever is superhuman is divine.

14.2 Eris – Goddess of Discord

This state of affairs is even clearer with the goddess Eris, who accompanies Phobos; for here, Homer precisely describes what happens as the armies march towards each other, reporting the following about Eris, 'Ares' murderous sister':

> Small when her crest first appears,
>
> But so ravenous and relentless in her ways
>
> That she soon thrusts her head into the sky
>
> Even while she keeps her feet on the earth.
>
> This horror now infused equal parts of strife
>
> Into both armies as she patrolled their ranks,
>
> Swelling the volume of human suffering.[5]

As long as the armies are far apart, the discord that caused them to march against each other still has its normal shape: its crest is still low – literally, 'it only has a low helmet' (*koryssetai*). It is implied that the crest is one of foam, as Homer uses the same word to describe the wave crests. Here the image has a twofold purpose: on the one hand, it evokes the notion of discord sweeping the men away like a flood. On the other hand, it anticipates the next astounding image, that of Eris's raised head. For when the armies are close to each other, the flood is no longer as low as it initially was; she thrusts her head up to the sky, even though her feet remain on the earth. This depicts that unbearable suspense from which Phobos erupts. For the armies, directly before battle starts, the entire world – that is the meaning of the image – is filled with the incredible power of Eris, and now she casts strife into their midst: the battle breaks out. Presumably you know the legend in which Eris, because she was

[5] Homer, *Iliad* IV, 442–445, p. 78.

not invited to the wedding of Peleus and Thetis, threw a golden apple among the three goddesses Hera, Athena and Aphrodite, who began a quarrel over it that was settled by the judgement of Paris. Just as Eris threw the apple in the tale, she throws strife itself among the armies. We can also become aware of the peculiar factual logic of mythical thinking through this image. Eris 'infused equal parts of strife into both armies'. A quarrel is not something that can be assigned to one partner or the other; it belongs equally to both. When we say that a quarrel 'rises up' between two groups, we use the same image with which Homer describes Eris's ascent to the heavens. But Homer is more precise: what rises up is in reality discord, and discord then casts strife among them.

Such a phenomenon cannot be described with our modern way of thinking, which knows only subjects and objects. When Homer presents it in the forms of mythical thinking, we can still visualize it today. Discord too is a superhuman power that seizes the armies. According to Hesiod, it is a daughter of baleful night and creates painful toil, neglect, starvation, pain, battles, combats, bloodshed and slaughter, quarrels, lies, pretences, arguments, disorder, disaster and perjury.[6] All these children of Eris are, like herself, superhuman powers that assail humanity. For purportedly rational thought, discord is an overarching concept under which the particular forms of discord are subsumed. But an overarching concept does not have the power to drive entire peoples to their doom, and the subsumption of particular concepts under an overarching one does not express the fact that the supposed sub-concepts are in fact the ghastly consequences of discord. But this is conveyed in the vivid image of torment, neglect, starvation, pain and so forth as the 'children' of Eris. Here we see once more that mythical thinking envisages the phenomena in their true form far more directly than our derived way of thinking, which is refracted through a multiplicity of prejudices. If we consider the example of Phobos and Eris, nothing gives us the right to claim that our rationality sees our conditions for what they truly are, while mythical thinking is blinded by superstition.

14.3 Affect and God

In the language of later philosophy, Phobos and Eris are affects. What is the difference between the divine powers in Homer and the affects? This question is easily answered: affects are processes within the soul. In the battle example, then, they are what goes on in the soul of each individual warrior. Now, no one would deny that processes take place inside each individual warrior that can, if one wishes, be viewed in isolation. If we do that, however, then the battle, the discord and suspense between the armies, and all that goes with it, disappears. We no longer have the true process in view, and can therefore no longer understand the processes in the soul either. For if the process itself disappears, then whatever is experienced in the soul

[6] Hesiod, *Theogony*, trans. Martin Litchfield West (Oxford and New York: Oxford University Press, 1999), p. 10.

processes and gives them their specific form in the first place will disappear along with it. Let me illustrate the overlooking that occurs here with an example. You can ostensibly describe the sounds of a symphony by Beethoven comprehensively, correctly and unambiguously by specifying the frequency of each individual note and listing these frequencies in the correct order. Then you will have, in the assemblage of all frequencies, the totality of the 'information' of which, from an information-theoretical perspective, this Beethoven symphony 'consists'. That would be precisely analogous to a description of the Homeric battle that used the sum of affects among the individual warriors to measure the overall discord and panic experienced. What is lost in such a description in the example of the Beethoven symphony is no less than the symphony itself, which takes place in its unity, its incredible tensions and releases and its architecture at a particular moment in a particular circle of listeners. This event is eliminated in the sum of frequencies, just as the sum of the warriors' affects eliminates the power that seizes and unsettles them. This power is something different from the affects, but the affects would not exist without this power. This power is *suffered* in the affects. The Greeks called such a power 'divine'. It is, when it bears a name, a god.

Theologians invented a particular word to describe gods like Phobos and Eris; they call these gods 'personificiations'. This is based on the idea that panic is 'in reality' an affect, and that the primitive thinking of the mythical age had not yet reached the level of rationality at which one can formulate an abstract concept to identify this affect. According to this theory, primitive peoples imagined general concepts gained through induction in the shape of persons, and referred to the concepts disguised in this way as 'gods'. I do not need to explain in how absurd a form the prejudices of a supposedly rational mindset, based on nothing, are concentrated in this. The actual historical process is precisely the reverse, as Karl Reinhardt has shown in his fine essay 'Personification and Allegory'[7]: the gods are not a product of personification; rather, every personification is preceded by the manifestation of a god. What we term 'personification', however, is only one particular form of what one could call conceptualization. I will therefore take the liberty of saying here in general that all conceptualisations presuppose the manifestation of gods. When we use such words as 'war', 'peace', 'poverty', 'wealth', 'beauty' or 'sickness, we squander the legacy of mythical thinking preserved in our language. Language, in all its forms and possibilities, emerged from mythical thinking. In language, mythical thinking is still alive today. Before the word 'terror' could be formed, the power that Homer calls 'Phobos' first had to be recognised. The term has no power and cannot refer to a power; it is abstract by definition. In the name 'Phobos', however, because it is a name and not a concept, this power is recognised and invoked. We can no longer control the powers we deal with, for we have forgotten that one can only refer to powers with names, not with conceptual terms.

---

[7] Karl Reinhardt, 'Personifikation und Allegorie', in *Vermächtnis der Antike*, ed. Carl Becker (Göttingen: Vandenhoeck und Ruprecht, 1960), pp. 5ff.

The lines from the *Iliad* gave us the first illustration of mythical thinking in this lecture. They therefore mark an important threshold in the systematic progress of our reflections. We must try to realise what can be learnt from them.

Phobos is – and this is beyond doubt – a god. Contrary to all the notions we tend to associate with the name 'god', however, this god – and this too is beyond doubt – is a phenomenon we can still experience today. Today, we imagine the same phenomenon in a different way; we let it encounter us in a different way. But it is possible to ascertain whether the Homeric or the modern way of envisaging the phenomenon is the appropriate one.

Today, Phobos is considered an affect rather than a god. We take this for granted, although this interpretation of the experience of terror is not an achievement of the Modern Age. We adopt it from tradition, reproducing a way of thinking that developed among the Greeks from the fifth century BCE and was given what would remain its definitive form by Plato, who was still able to look in both directions. Affects are processes within the soul. So we see here a first example of how the region known since Plato as the 'soul' came into being through a transposition of divine powers to the inner life of humans. Because of this turn of the consciousness away from the forms of mythical world experience, the word 'god' was destined to become incomprehensible within a short time. But the intellectual landscape that we call 'Europe' arose at the same time as the disappearance of the gods and the discovery of the soul.

But Phobos is also central to the field of aesthetics. Aristotle, in his famous definition of tragedy (*Poetics*, 1449b),[8] states that it achieves a purification of affects through Phobos and Eleos, fear and pity. I will not deal with the interpretation of these oft-debated words for now; anyone who is interested in this should consult the important, but far from exhaustive essay 'Fear and Pity?' by Wolfgang Schadewaldt.[9] In our context, all that is important for now is that the background to this definition is an earlier theory about the effect of *music*. The power of Phobos is inseparably tied to aural perception. Phobos and Eleos, fear and pity, should be viewed as a unity; this mirrors the ambivalence in the nature of Phobos that already became clear in the *Iliad*. The artistic representation of processes that invoke this power causes an unsettling (*kinēsis*) of the soul, which enters a state referred to by the Greeks as the 'calm of the sea' (*galēnē*). This calm would later become Kant's 'disinterested pleasure'. In our systematic line of reasoning, this unsettling of the soul, which is here caused by Phobos, but can also be brought about by other powers, appeared as the vibration of our sensual sphere in its totality; we owe our capacity for *perception* to it. May this suffice to show that one and the same phenomenon, Phobos, reveals itself under four different aspects: it appears as a god, as an affect, as a fundamental determination of aesthetic feeling and as a condition of the possibility of perception. The modern consciousness tears these aspects apart. Then they appear disconnected in theology,

---

[8] Aristotle, *Poetics*, trans. Anthony Kenny (Oxford and New York: Oxford University Press, 2013), p. 31.

[9] Wolfgang Schadewaldt, 'Furcht und Mitleid?', in *Hellas und Hesperien*, vol. 1 (Zurich: Artemis, 1970), pp. 194–236.

psychology, aesthetics and the theory of perception. This breaks up both the unity of the phenomenon and the unity of the world from which it emerged. Our systematic line of reasoning is meant to uncover the horizon in which these fragments can be brought together once more.

14.4 The Invisible Power of Horror

It would be possible to explicate the example from Homer even more precisely in all these directions. It would then emerge that the 'theological' aspect of Phobos, whose existence is scarcely known to anyone today, is the only one that encompasses all dimensions of the phenomenon 'terror'. It is not merely an aspect; it provides the whole, and does so in its worldliness. In post-metaphysical views of the phenomenon, on the other hand, the phenomenon is particularized and broken out of its relations to the world. This is connected to the appropriateness of mythical thinking, which I illustrated earlier with the Homer example. I will not yet go further into all this, however, but only draw your attention to one particular element of this description that advances our analysis of listening. We had noted that the sensory perception of listening differs from the sensory perception of seeing in that the ear perceives powers, not objects. In the perception of powers, I would argue, unlike in the observation of objects, the distinction between inside and outside disappears. This too can be ascertained using the example of the god Phobos. The horror that arises between the armies is a phenomenon that we cannot locate at one particular point. If it were only in the centre between the armies, neither the Trojans nor the Greeks would be horrored. If it were only in the hearts of the Trojans, or those of the Greeks, both parties would not be threatened from within at the same time. In reality, this horror is a phenomenon that both arises in the centre between the armies and can suddenly appear within each individual warrior when he is seized by horror. But the rage of battle and danger are also a game with the all-pervasive power of horror. The Spartans sacrifice to Phobos so that horror will break out from their threat and attack their enemies. You see: there is no inside and outside here, for the power of horror does away with this distinction. This is also expressed in the German verb *entsetzen*, which literally means 'raise from one's seat or rest'; accordingly, *erschrecken*, 'startle', literally means 'cause to leap up'. We find our position of calm in the world, which shows us observation through the eye; for observation places objects at a distance from us, and we call everything that is distanced from us 'outside'. When powers appear, however, we realize that this geometric order of distances in which we seek our security is an illusion. The powers not only push against us; they infuse us, and we feel like their playthings. They make us feel as if we had been displaced from the visible world, where we stand on solid ground, to an invisible sphere that has no foundation, no 'seat', but is rather baseless – we are 'unseated' [*entsetzt*]. If one pays attention, one can ascertain that every perception of energy has this effect. But the technological world is so accustomed to dealing with enormous energy potentials that we now only acknowledge, but do not experience, what people in earlier times felt during the blowing of the wind, the roaring of the sea, during storms or earthquakes. Today it takes something in the manner of radioactive radiation to make us feel some of that horror, which is, in a sense, the aura of human existence.

14.5 The Perception of the Invisible

Yet even if we numb ourselves to it, this aura of perception through listening is always disclosed. We hear what is invisible – for powers and energies are invisible (I will speak later about light). The gods are also invisible, and we behave towards them like children who say that whatever cannot be seen is not true. In the field of science, this would be analogous to saying that whatever cannot be objectified is not true; that whatever cannot be quantified does not exist. Whether among scientists or children, this stems from the fact that the invisible makes us uneasy. The ancient Greeks often speak of the invisibility of the gods. The gods reveal themselves through their acts or by signs; if they wish to show themselves, they hide in a foreign guise, that of a human or an animal; the visible guise is then the form in which the gods make themselves *unrecognizable*. It is when one reads what the Greek poets write in clear words that one realizes how absurd all the talk is about the anthropomorphic form of the Greek gods – even if this talk came about among the Greeks themselves. Admittedly there is such a thing as 'seeing' the gods; but this is of a very different nature from observation with the human eye. Homer says: 'Gods are daunting when they appear as they are'.[10] In Book V of the *Iliad*, Athena frees Diomedes's eyes from the mist that had shrouded them; now he can recognize the gods, while they remain invisible to all other warriors.[11] Poets are the kin of prophets and seers, for they too – like Diomedes in this extraordinary moment – can recognize the gods and their works, but for this they require divine inspiration from the muses. If they wish to say what they see, they must translate their stories back into the forms of intuition that the clouded eye of the other humans can process. It was from this power of sight conveyed by the muses that early philosophy derived its capacity for non-sensual intuition; this is why Plato dedicated the Academy to the cult of the muses and termed philosophy the highest art of the muses. We still have to deal with this 'seeing of the invisible', for mythical figures and art both originate from it. The first concern was to show that through hearing, we constantly perceive the *reality* of this invisible sphere. We must deny aural perception if we are to claim that only what is visible is real.

This is where I can finally respond to an objection I had promised I would address. The statement that we perceive energy by hearing, not through the eye, was countered by the observation that we see light. In the context of this lecture, it is legitimate to let a poet reply to this first. In Goethe's 'Pandora', Prometheus says that the human race is 'destined to see what is illuminated, not the light.' In the prelude to Act One of

[10] Homer, *Iliad* XX, 131, p. 391.

[11] *Iliad* V, 127f., p. 87.

Faust II, Goethe combined this notion, which should be read against the background of Goethe's *Theory of Colours*, with the interpretation of all human striving that forms the basic theme of the work. In a long soliloquy, Faust describes the dawning of the day and the rising of the sun. First the light strikes 'the vast mountain heights' (4695) and then gradually turns past the 'sloping Alpine meadows bright' (4699) and down towards us:

And step by step it reaches us down here: –
It blazes out! – Ah, already blinded, though
I turn away, my eyesight wounded, pierced.

So it is, when to the thing we yearn for
The highest wish so intimately rehearsed,
We find fulfilment opening wide the door:
And then, from eternal space, there breaks
A flood of flame, we stand amazed before:
We wished to set the torch of life ablaze,
A sea of fire consumes us, and such fire!
Love, is it, then? Or hate? This fierce embrace,
The joy and pain of alternating pyres,
So that, gazing back to earth again,
We seek to veil ourselves in youth's desire.

Let the sun shine on, behind me, then!
The waterfall that splits the cliffs' broad edge,
I gaze at with a growing pleasure, when
A thousand torrents plunge from ledge to ledge,
And still a thousand more pour down that stair,
Spraying the bright foam skywards from their beds.
And in lone splendour, through the tumult there,
The rainbow's arch of colour, bending brightly,
Is clearly marked, and then dissolved in air,
Around it the cool showers, falling lightly.
There the efforts of mankind they mirror.

Reflect on it, you'll understand precisely:

We live our life amongst refracted colour. (4701–4727)[12]

'Reflect on it' – if we wanted to follow this exhortation, we would have to go deep into the sunken kingdom that the name 'Goethe' did not disclose to the educated classes and their current epigones, but rather hid from them. I will simply point out that the idea that we are 'destined to see what is illuminated, not the light' appears here on several levels at once. First we hear of the sun's light, and are told that our eye cannot see this light directly, only in the prismatic refraction of its colourful image in the waterfall. This is an optical observation that rests on Goethe's massive work the *Theory of Colours*. Our eye never sees the light itself; it only sees what is illuminated *in* the light. If one describes light not in the manner of Goethe but of Newton, it reinforces the seemingly paradoxical but scarcely refutable statement that light as such is invisible. We see what is illuminated, and if we refuse to be dissuaded from our belief that we are seeing the light, it is because the illuminated is a 'phenomenon', because it is transparent for the light in which it appears. Thus we gaze indirectly and refractedly through the illuminated phenomenon into eternal expanses, from which, as Goethe puts it, overpowering flames erupt; we recognize the world in every phenomenon by the reflection of light.

In Goethe's poem, however, the rising sun and the waterfall are themselves 'refracted colour', namely an image – or, more precisely, a metaphor. This is already evident in the description of the light:

A sea of fire consumes us, and such fire!

Love, is it, then? Or hate? This fierce embrace,

The joy and pain of alternating pyres

Here the fiery sea of light is a metaphor for the fiery sea of life, which is experienced not as a private or finite, let alone bourgeois existence, but as a universal, divine, world-creating power. The universality and divinity of life, in which our existence is no more than a drop of foam in a waterfall, emerges in the metaphor through the fact that the universality of light, described so powerfully, mirrors the power of life – that the relation between light and life equals that between the waterfall and light. The notion that the eternal creative power of endless life is prismatically refracted in all manifestations of nature is the basic idea underlying Goethe's writings on morphology, that is, on geology, comparative anatomy and botany. But the same conception also defines Goethe's ideas on the theory of art; with the reference to 'refracted colour', Faust's soliloquy also contains his interpretation of his own poetic work.

Let us return to the light. Faust's soliloquy teaches us to understand that humans are destined to see what is illuminated, not the light. The true nature of light can be perceived by divine beings. Goethe prefaced the soliloquy with an exchange between

---

[12] Johann Wolfgang von Goethe, *Faust Parts I & II*, trans. A. S. Kline, www.http://www.poetryintranslation.com, accessed 15 November 2019.

Ariel and his spirit choir that shows us how the same sunrise is witnessed by Ariel. Spirits, we learn, do not perceive the true and unbroken nature of light through the eye, but by hearing:

> Listen! Hear the Hours nearing!
>
> Ringing out to spirit-hearing,
>
> Now, the new day is appearing.
>
> Doors of stone creak and chatter,
>
> Phoebus' wheels roll and clatter,
>
> What a din the daylight's bringing!
>
> Trombone- and trumpeting,
>
> Eyes amazed, and ears ringing,
>
> The Unheard drops out of hearing. (4666–4674)

This is indeed an 'unheard' truth, for it is inaudible to humans: that the true, energetic nature of unbroken light manifests itself to the spirits acoustically, not visually, as an incredible racket. To make this idea tangible, Goethe has disrupted the flowing, mellifluous 'naturalness' of his new lyrical language and, in an act of conscious defamiliarization, mobilized the entire apparatus of gods and words available in Baroque theatre and poetry to let us hear the grinding of machines: the Hours (Horae), Phoebus, the chattering creak of stone doors or the sounds of trombones and trumpets. The language simultaneously gives directions to the composer who would set this song to music. In Baroque poetry, the droning bombast of these artistic devices strikes us – probably unfairly – as naive. But when Goethe resorts to such devices at the pinnacle of his fully-developed poetic art, the intended result, far from any naiveté, is something very different, something typically modern: the Classical and Biedermeier periods are brushed aside along with the world of sentimentalism, and behind them emerge, in conscious dissonance with the spirit and taste of the age, the monumental foreignness and sacred pomp of the sixteenth and seventeenth centuries – but with an unmistakable irony that forbids us from confusing the description with the things it describes. Aesthetically, this creates such amazement that the listener is enabled to hear the sense of the words: 'The Unheard drops out of hearing'.

I have taken a certain amount of time to present this poetic source for the claim that energies and powers are perceived through the hearing, not the eye, in order to let poetry have its say in this lecture about art with an astonishing example. We will now return to our topic and attempt further to elucidate the nature of the affects.

14.6 Affect and Anticipation

I have explained the difference, or rather, the opposition between the Greek term *pathos* and the Latin term *affectus*, which has suppressed and covered up the experience expressed in the word *pathos*. *Pathos*, as we have seen, means 'something experienced'. To clarify the nature of such experience, I picked out the example of an 'affect' that is especially significant in the philosophical theory of affects and in aesthetics: Phobos. We will now consider, independently of the myth, what happens to humans here and *how* it happens to them.

Contrary to the form in which, according to the usual view, sensory organs are affected by impressions, that is to say, contrary to the so-called affections of the senses, we find that with fear and horror, there is no object that could affect the senses. One is afraid because one *expects* something terrible. If what is feared occurs, one can feel pain or despair – but the fear has vanished; it has lost its purpose. Fear is directed not at the present but at what is to come; it is an anticipation of the future. There have been manifold attempts to classify the affects, even though they consistently have a trait that defies all classification: they are consistently ambivalent. We saw this in the case of Phobos, who has the twofold face of horror and threat. This ambivalence explains why, since Plato, all attempts to classify the affects have traced them back to the binary opposites of pleasure and pain (*hedonē* and *lypē*). Then one can derive one class of affects from the basic form of desire (*epithymia*), the other from the basic form of flight. Both striving for pleasure and fleeing from displeasure can be followed through our entire sphere of sensuality. The specifications then result from the differentiations within this sphere and the differentiations among the so-called drives. In this manner, one can derive a complete schema of classifications and arrive at a large, clear chart of affects. The usefulness of such charts is beyond doubt; nonetheless, their underlying principle is unfortunately wrong. For pleasure and pain are themselves ambivalent; the one turns into the other. There is no pleasure without pain, no pain without pleasure. (Anyone who would protest against the second of these claims should learn from psychoanalysis; they will acquaint themselves not only with the polyvalence of pain, but also with the reasons for their protest.) This ambivalence running through everything can be shown in every single one of the so-called affects. It is one of the most basic facts of all anthropology. In the *Philebus*, Plato developed both the clearest and the most profound formulation of his dialectics from an examination of the ambivalence of pleasure. The nature of humans in the entire domain of the affects, that is, in the whole sphere of their sensuality, is such that wherever they strive, they simultaneously flee, and wherever they flee, they simultaneously strive. If fleeing and striving are the basic forms of the affects, this also means that what we have seen in the example of fear applies to all affects: they refer to what is not yet there, to what is not yet present. They are all anticipations of a possible future. In desire, humans strive to appropriate future possibilities. In flight and defence, they shrink back from such possibilities. Hence we can make the general statement that the realm of affects is a realm of possibilities.

Chapter 15
Epilogue: On Evil (1981)

Evil is the unthinkable as such.[1] It is all too familiar, certainly; our conscience tells us through warning signals that it is covertly active within us. But we cannot take hold of it; it dances around us wearing a thousand masks. It not only reveals itself in misdeeds and crimes, but also delights in hiding within our morals, our ideals, in successes, glorious deeds and achievements. Where it emerges undisguised, we gaze into an abyss. Even its masks emanate a horror that makes us shrink back and flee. It is precisely in this flight, however, that we are bound by its spell. Evil is a *fascinosum*, and perhaps everything that 'fascinates' us mirrors the temptations of evil. It even appears to us in the guise of beauty; then it fascinates us with its hidden coldness and through narcissistic reflexes. Because evil is able to don any mask, we have a tendency to tell ourselves that we can defeat it by using the critical scalpels of the masters of unmasking. Thus Darwin, Marx, Nietzsche and Freud have been called to our aid as exorcists of our civilization. But unmasking is itself an evil act; it reveals, but does not build up. As a result, we stand here today with empty hands. We have forgotten the ancient wisdom that evil can only be defeated by good. If we are asked what good is, we find ourselves at a loss; we have no answer, and usually seek refuge once more in one of evil's masks.

The Christian tradition in which we grew up—which has moved far away from the Gospel—sought out evil in the soul of the individual; it called it 'sin'. But the history of our century has taught everyone who did not already know that evil resides not in the individual but the collective consciousness. It is in the collective consciousness that the clouds gather from which the thunderbolt of evil strikes. We are evil because we allow the collective consciousness to enter our own consciousness without resistance. The space we call our 'soul' is then revealed as a vacuum that sucks up what we ourselves cannot espouse. Wherever such a vacuum exists in cultures and societies, in groups and individuals, evil finds its lair. It would be harmless if we only

[1] This text was first published as: "Über das Böse". In: *Hier und Jetzt* Vol. II, p. 484–500. The permission to include this text here was granted by Klett-Cotta.

E. Rudolph and J. Picht (eds.), *Georg Picht: A Pioneer in Philosophy, Politics and the Arts*, Pioneers in Arts, Humanities, Science, Engineering, Practice 19,
https://doi.org/10.1007/978-3-030-31790-4_15

encountered it in criminals and miscreants. But the soil in which evil thrives consists of the shallow minds and indifferent hearts—mediocre, the narrow-minded, the functionaries, the manipulable masses in all strata of society. The machines of terror are operated not by monsters but by grey, paltry ordinary humans. Wars of extermination are waged by automata of an abstract sense of duty. The technicians of genocide are not beasts but everyday people. Once the horror has taken place, everyone will claim they knew nothing and were not involved. And that is actually true, for it was not the person, but the vacuum in the person that allowed itself to be filled by the collective evil. The person was already so hollowed out that they could not perceive what happened to them; they dance like a cork moving in the waves of a collective consciousness. No individual can be held accountable for the facts that caused the societies of the twentieth century to be driven so thoroughly by the power of evil. The current conditions can be explained by a naïve ignorance in the face of 'so-called' evil[2] that has a long history, and whose symptoms are manifest in the global crisis of scientific-technological civilization. This would make evil a central issue for both politics and philosophy; we would have to comprehend evil in order to fight it. But, as we said, it is the unthinkable as such. Evil shatters all categories.

We cannot think evil, but it has power over our thoughts. We therefore cannot but search for criteria that might signal to us where evil infiltrates our thinking. Here one can make a very simple observation: evil is most powerful where one refuses to acknowledge it. In so doing, one deprives oneself of the ability to perceive reality as it is. This applies, for example, to a science that prides itself on being 'value-free'. It applies to the translation of its abstractions into technical, industrial and social mechanisms. It applies to every 'system' whose functional stability rests on the fact that the individuals who operate it are interchangeable; their aptitude is proved by the fact that they react to the indifference of those systems with an indifference of their own. This blindness of our time is characterized by the fact that even after Auschwitz and Hiroshima, economic theory, the social sciences and the so-called humanities either completely omit the issue of evil or, if it does appear, do not make it an object of methodological reflection. In analytic philosophy and structuralism too, evil is unknown. It is subject to a collective taboo. That the same individuals who practise these disciplines express outrage at the evil found in the world today, yet are often inclined to devote themselves uncritically to some ideology, does not have any impact on their methodological awareness. This schizophrenic behaviour is a symptom of the all-pervasive power of evil. But how can we recognize evil if we are unable to think it? How can we say what is at issue if evil eludes conceptualization, or if it even transpired that the concept as such is already a manifestation of evil?[3]

Religions and cultures that, unlike us, knew the power of evil, knew that one can only defeat it by binding it. Evil is bound in shapes and through rites that delimit an inner or outer zone of protection. It is not bound by analysing it, but by almost invoking it, forcing it to reveal itself in unconcealed form. European philosophy

---

[2] A reference to Konrad Lorenz's book *Das sogenannte Böse* (Vienna: G. Borotha-Schoeler, 1968) (ed.).

[3] See 'Philosophy and Politics', pp. 123–129.

barely attempted this between Plato and Nietzsche, who both bring evil to manifestation. A representation that binds contradicts the perspective and rules of the craft that came into force after Aristotle. We will discuss the two great exceptions, Schelling and Kant (for Kierkegaard speaks of sin as a theologian), later on. We will also have to ask why metaphysics knows nothing of evil, and perhaps did not want to know anything. If binding is the form in which we recognize evil, it becomes clear that the knowledge of evil in European culture remains limited to those spheres of the human spirit that have not placed a taboo on the corresponding forms of representation. Evil is one of the great themes of art: sculpture, painting, poetry and literature, and—with certain restrictions—also music. Here it was Wagner who first broke through the protective circle with which music had previously, in the powerful tradition of *musica sacra*, been able to shield itself from evil. Through Wagner it learnt to unleash demons, and thus open up a field that had long been familiar in the music of other cultures. Representation, as I hope to show elsewhere, is an elemental, in fact the primary form of all cognition. If we concede that we think in every form of cognition, we will have to change the premise from which I began: European art and literature made evil thinkable; European philosophy and science resisted what was thought here by dogmatizing a concept of thinking that excluded evil from the realm of the thinkable. If evil is a true power, however, then philosophy and science, through this choice, deprived themselves of the possibility to recognize reality as it is, namely in its truth.

Misunderstood religious and philosophical traditions in Europe have led to the preconception that evil is a privilege of humans, while extra-human nature is good. Humans even lay claim to a monopoly on the absolute power of evil. This cannot be reconciled either with our direct experience of the world or the insights of modern natural science. All living beings survive by destroying or pushing aside other beings; this basic law, which is part of the very definition of life, is implicit in the principles of evolution and selection.[4] The origin of evil in all its forms lies in life itself. Whoever refuses to accept this has already surrendered to self-deception, and thus to falsehood. They cannot overcome evil because they cannot bear the sight of it, which means that they have already succumbed to it. If the origin of evil lies in life itself, we can only live truthfully if we are willing to acknowledge it. Powerful and awe-inspiring traditions of thought, not only in Europe, have been led by this insight to deny life as such. But if we deny life, good disappears; for what we call 'good' arises from the affirmation of life. Based on this, we can formulate a principle that unsettles all our prejudices about good and evil: good can only unfold if we acknowledge evil. This is different from, and more than, a resigned observation that evil cannot be eliminated from the world. It means that overcoming evil begins not with its denial but when we accept it and take it upon ourselves as one of the conditions of life as such. Good and evil cannot be placed in opposition according to the principle of contradiction; both are equally originary components of life, and each is in constant transition towards the other. There is no good that is entirely devoid of evil, and no evil that is entirely devoid of good. The words 'good' and 'evil', like the words 'light' and 'night', refer

[4] See 'Philosophie und Völkerrecht', in: *Hier und Jetzt* Vol. I p. 57–115.

to extremes in a prism of infinite transitions. All living things appear in this prism; therefore, all living things are ambivalent. This is connected to the fact that all living things are in time.

Humans are products of their evolution. If we examine the manifestations of evil in the different spheres of human existence, then we can, as behavioural science has shown, develop a morphology of evil that traces almost every element of human life that we call 'evil' to patterns of behaviour that were passed on to humans as an indelible legacy from earlier stages of their evolution. The human imagination cannot produce anything whose blueprint is not prefigured in the collective memory of our genetic inheritance. If we occasionally recognize the beasts that reside within us in our own behaviour or that of others, in dreams, myths or works of art, we are startled and say, 'I don't recognize myself in that'. But this self also includes the jungle that each of us conceals within ourselves as an indelible memory of our genetic background. This genetic background is common to all humans; the memory of the jungle is not individual but collective consciousness. What we call the 'soul' is not primarily our own self, but rather collective memory. The collective consciousness is the condition of possibility for individuality; it always takes precedence over individual consciousness. If our 'soul' is collective memory, the word 'I' contains a self-deception that negates life. Through this word we lay claim to a monopoly to which we are not entitled: a monopoly on the possession of our 'own' soul. But we cannot erase the fact that we are humans by virtue of our souls, our share in the collective memory of our genetic inheritance; whoever stifles their soul becomes evil. Therefore, we must not deny the jungle within us, but would do well to accept it calmly, just as we must accept nature as it is. There is no emancipation from our genetic origins. For reasons we will go into later, humans have the possibility to set themselves the task of overcoming these origins. But we will always fail at this task as long as we deny them. Whoever seeks to fulfil the task they have been set within the prism between good and evil will first have to accept who they are.

It is not an unacceptable anthropomorphism if we recognize evil as we know it from our own soul in the nature that surrounds us, then, but rather a transparent result of evolution. It is anthropocentric thinking to imagine that cruelty, violence, malice, deviousness or deceit are only evil when we encounter them in humans, whereas these attributes are worthy of praise in animals because they have developed in evolution as conditions for the survival of their species. We humans overtook the other animals in the selection process because the same attributes were developed with greater efficiency in us. Our success in the 'struggle for existence' does not take us out of nature, but rather into it. According to the bible (Genesis 2:8 ff.), man only steps out of nature by eating from the tree of the *knowledge* of good and evil. Thus he distances himself from life. He recognizes that the meaning of the words 'good' and 'bad' is not affected by whether the one or the other behaviour has a greater chance of winning out in the course of evolution. This only makes the true nature of evil more enigmatic.

We cannot say that success in the struggle for existence is good and what is removed via selection is bad, nor can we conversely assert that whatever succeeds must therefore be evil. But nothing entitles us to the opinion that humans can take possession of good and evil by recognizing them, that only humans can be good or evil. The anthropomorphic interpretation of 'good' and 'evil' that established itself in the Christian tradition does not chime with the account of the creation. It was imposed on this text because modern moral philosophy teaches that only someone who knowingly does evil should be called 'evil'. This claim was used by the Germans after 1945 to evade their collective guilt for Auschwitz. Because we have seen through the deceitfulness of the individual and collective unconscious with its mechanisms of repression, no one can invoke such a doctrine with a clear conscience. Compared to the self-deception of the masses who carry out evil at the same time as pretending they neither know nor want it, the honesty of undisguised evil seems like innocence.

The fact that man ate from the tree of the knowledge of good and evil does not prevent, but rather enables to recognize in nature what we find in ourselves as evil. Mythical thinking, which encounters evil in the form of terrifying animal forms, is more deeply embedded in reality than the supposedly enlightened abstractions with which we attempt to explain away evil from nature so that we can repress it in ourselves too. What follows from this is a conclusion that contradicts all our preconceptions, namely that 'evil' is not a moral concept. Humans could not have morality if evil did not exist in nature. The origin of both evil and good in life itself is the condition of possibility for morality; morality is not the condition of possibility for good and evil. We cannot even say that morality is good because it sets the rules by which we normally distinguish between good and evil, for morality, by excluding evil from itself, is always at risk of becoming one of its strongholds.

The principles of evolution and selection regulate the self-preservation and species preservation of organisms. We are accustomed to thinking that whatever is conducive to self-preservation and the preservation of species is positive and whatever harms it is negative, and we are reluctant to consider that from this perspective, the competition between organisms presents as many contradictions as there are competing species and individuals. From the perspective of the other living beings, the human being of our civilization can only seem like the incarnation of evil, for it disposes over them as it pleases, exterminates them and destroys their ecological niches. Even those who preach freedom and non-violence nourish themselves by devouring other organisms, and have no concerns about poisoning bacteria or viruses when they become ill.

If whatever is conducive to self-preservation is considered good, and if the self views itself as an 'I', that is, as a subject that dominates nature, obeys no higher law and occupies a position above and outside nature, the only possible result is destruction. Leibniz (in the tradition of Greek philosophy) sought to solve this problem with the grand idea of the harmony of the world. If one presupposes the harmony of the world and the concept of God that underpins it, one can claim that whatever happens is necessary for the preservation of the whole. It may appear to be damaging from the perspective of an individual, a group or a species, but this only applies in relation to their particular location. The 'self' that must be preserved is not the respectively individual element that exists in the world, but rather the order of the cosmos as a whole, whose continued existence is the prerequisite for everything individual. Evil disappears in this grand metaphysical conception; it dissolves into individual perspectives.

Through Leibniz, however, metaphysics takes the form of theodicy; he coined this term.

> By theodicies one means defences of the highest wisdom of the Creator against the complaints which reason makes by pointing to the existence of things in the world which contradict the wise purpose. – One calls such a defence a 'plea for God's cause', although it is really no more than the cause of our presumptuous reason, which fails to recognize its bounds.[5]

The self-elevation of man to the judge over God implies an inversion of good and evil. That is why Luther already radically rejected the speculative questioning that Leibniz later termed 'theodicy': it is not God who must justify himself to the forum of human understanding, but humans who must be justified before God. The heart of Kant's essay, which begins with the lines quoted above, is an extensive interpretation of the Book of Job. It leads him to expand his criticism of Leibniz into a withering critique of speculative theology as such:

> Job probably would have met with a sorry fate if faced with any court of dogmatic theologians, a synod, an inquisition or an ecclesiastical convocation of our time (with a single exception).[6]

The idea that God should have to justify himself before the judgement seat of human reason is a precise inversion of the idea of the Day of Judgement. It is the utmost blasphemy of which human thought is capable. Theodicy is therefore a manifestation of evil.

As stated above, Kant, despite his attack on dogmatic theology (or indeed because of it), stands in a clearly Lutheran tradition. But his criticism of Leibniz was presumably triggered by a modern spirit that cannot be suspected of Lutheranism. Voltaire reacted to the philosophy of Leibniz with an outburst of elemental rage that shook the entire fabric of the political and intellectual world of his time. His satirical novel *Candide* is a savage mockery of Leibniz's doctrine that the world as God created it is the best of all possible worlds. Here it is not faith, as with Luther, but reason that protests against theodicy. The European Enlightenment only succumbed to Leibniz's optimism later on; it transformed it into the idea of progress. Hegel concludes his *Lectures on the Philosophy of History* with the following words:

> That the history of the world, with all the changing scenes which its annals present, is this process of development and the realization of spirit – this is the true *theodicy*, the justification of God in history.[7]

One cannot take the denial of evil further than this. When, by contrast, Nietzsche dedicated *Human, All Too Human* 'To Voltaire's memory, in commemoration of the day of his death', he did so to call the authentic form of enlightenment to mind once more: enlightenment is the maturity of a spirit that proves its freedom by not closing its eyes to evil.

---

[5] Immanuel Kant, 'On the Failure of all Attempted Philosophical Theodicies', in Michel Desplant, *Kant on History and Religion* (Montreal: McGill-Queen's University Press, 1973), p. 283.

[6] Ibid.

[7] G. W. F. Hegel, *The Philosophy of History*, trans. J. Sibree (New York: Dover, 1956), p. 457.

Immanuel Kant was one such spirit. However, he stands in the tradition from which we turned away above: he locates good and evil purely in humans. The first part of his text 'Religion Within the Boundaries of Mere Reason' bears the heading 'Of Radical Evil in Human Nature'. According to Kant, the nature of humans lies in the faculty of reason; the condition of possibility for reason as such is its faculty of self-determination: freedom. Freedom is the capacity for good and evil. Therefore, the doctrine of radical evil in human nature states that evil is not the only condition of possibility for freedom, but certainly an ineluctable one. Evil cannot be grounded merely in a natural impulse, for then

> the entire exercise of freedom could be traced back to a determination through natural causes – and this would contradict freedom. Whenever we therefore say, 'The human being is by nature good', or, 'He is by nature evil', this only means that he holds within himself a first ground (to us inscrutable) for the adoption of good or evil (unlawful) maxims, and that he holds the ground *qua* human, universally – in such a way, therefore, that by his maxims he expresses at the same time the character of his species.[8]

We cannot embark here on an interpretation of Kant's doctrine of radical evil and its connection to the doctrine of transcendental semblance, and will only note that this doctrine heralded the great crisis of European metaphysics, for metaphysics is a way of thinking that reveals its untruth through its inability to recognize and acknowledge the reality of evil.

After Hegel undertook the restoration of metaphysics in the *Phenomenology of Spirit*, Schelling countered him in 1809 with his treatise *Philosophical Investigations into the Essence of Human Freedom*. I consider this text, alongside Heidegger, to be one of the greatest documents of European history. While returning to Kant, it makes a superhuman attempt to think freedom absolutely as the capacity for good and bad and to derive them, both good and bad, from the nature of God. God himself is the first ground of good and evil, termed 'inscrutable' by Kant. For Schelling too, the reality of good and evil only comes to light in the human being. But because he thinks both in absolute terms—that is, in God—he is able to construct the entire philosophy of nature as a prehistory of the manifestation of good and evil. This brings the human being back into nature. Yet Schelling's treatise remains metaphysics; in fact, it is even a theodicy. Hence the failure of his attempt to refute Hegel's cancellation of evil. He looks ahead to a new horizon of thought, but cannot reach it himself.

Through their relative freedom and a qualitative evolutionary leap, humans indeed gained a special position in relation to other living beings. Kant interpreted this relative freedom absolutely: it separated human beings from nature and granted only to them the capacity for good and evil. Schelling, as we saw, brings humans back into nature, but he thinks this is only possible by doing away with Kant's delineation of the boundaries of human knowledge and thinking nature, like freedom, as the appearance of the absolute. If this can be recognized and evil can be thought, then along with freedom, knowledge too can be posited absolutely. Schelling asserts that the system of

[8] Immanuel Kant, 'Religion Within the Boundaries of Mere Reason', in *Religion Within the Boundaries of Mere Reason and Other Writings*, ed. and trans. Allen Wood and George di Giovanni (Cambridge University Press, 1998), p. 47.

true metaphysical knowledge is absolutely identical to the system of the world.[9] But this means that the spirit of the human being attains absolute identity with the spirit of God. Through this step in the idea, the human previously brought back into nature is broken out of it again; now this human even claims unconditional knowledge of the absolute truth in the theoretical use of reason. In both Kant and Schelling, it transpires, the rupture comes about through the idea of the absolute, which is thought as absolute identity. The philosophy of German Idealism took exception to the fact that in Kant, through the combination of the finitude of theoretical cognition and the unconditionality of practical cognition, a crack of sorts appears in identity. It tried to restore this identity without abandoning Kant's discoveries; this failed. The question arises whether we should perhaps proceed from Kant in the opposite direction and also question the absolute character of human freedom. Is the claim to the unconditional, for example, the idea of absolute identity, in every form a manifestation of evil, a deceptive fulfilment of Satan's promise: *eritis sicut Deus*—'ye shall be as gods' (Gen. 3,5)? Were humans delivered to the dominion of evil by the idea of unconditional freedom, which provides them with an unconditional power to control nature, in the very place where they believed they had discovered the origin of all morality?

In asking these questions, we are relying on Kant precisely where we are moving away from him. According to Kant, the nature of the human being manifests itself in the fact that, unlike other living beings, it is conscious of itself; it is expressed in the ability to say 'I'. Kant defined the being-'I' of humans with the Cartesian formula 'the *I think*': 'The *I think* must *be able* to accompany all my representations'.[10] But he interprets this formula differently from Descartes. As he shows, it has the logically paradoxically form that the 'I think' is neither a concept nor a judgement nor a conclusion. According to Kant, it has the 'logical form' of the 'problematic concept'.[11] The word 'problematic' does not mean that such a concept is a questionable matter; rather, Kant is adopting the parlance of mathematicians. In mathematics, a 'problem' is a task that is posed. Accordingly, for Kant, the 'I think' is only possible because it conceives of itself as an infinite task in its being-itself, that is, in its intended identity with itself. It must determine itself with a view to the idea of this infinite task in order to be capable of being and thinking. It 'is' only because it thus makes itself possible; in this way, it determines itself to itself. By virtue of its self-determination it is its own ground, suspended above a first ground that is inscrutable to us. Kant refers to this transcategorial ontological structure of the 'I think' as 'autonomy'. Autonomy is the formal structure of the faculty to be a cause through one's own spontaneity: of the will. Therefore, freedom is the condition of possibility for the 'I think' as such in both its practical and its theoretical usage.

---

[9] See F. W. J. Schelling, 'Stuttgart Seminars (1810)', in *Idealism and the Endgame of Theory: Three Essays by F. W. J. Schelling*, ed. and trans. Thomas Pfau (New York: SUNY Press, 1994), pp. 195–243.

[10] Kant, *Critique of Pure Reason*, p. 246.

[11] Ibid., p. 362; see p. 592.

When the nature of the I is developed in such a way, we can no longer succumb to the delusion that the word 'I' refers to the empirically existing individual in its individuality. Every time we say 'I', the empirical consciousness passes into the 'I think' of the transcendental consciousness, and the transcendental consciousness is identical in all empirical individuals. It is, to employ the phrase we used at the start, collective consciousness. The difference between Kant's transcendental and our collective consciousness is that Kant's transcendental consciousness is by definition true consciousness; if, on the other hand, we speak of collective consciousness, the relationship of this consciousness to the truth remains open. On the one hand, the difference between the two terms relates to the fact that we view the relationship between consciousness and time differently today. Kant's transcendental consciousness is timeless. On the other hand, it is impossible to overlook a difference of attitude towards evil in the two interpretations. In Kant, the (finite) truth of transcendental consciousness is guaranteed by the fact that reason is built as a system, and thus consistently carried by the identity towards which the 'I think' orients itself as the infinite task of its ground of determination. That is why Kant can claim that the cognition of practical reason is true a priori and therefore also good; this means that humans are a priori capable of distinguishing good from evil. If, by contrast, we speak of collective consciousness, we call into question the a priori nature of the knowledge of good and evil. We consider it possible that identity as such, even if it is thought by practical reason as a mere idea, is transcendental semblance, and that we are threatened by this semblance whenever we dare to use the word 'I'. This changes the constellation between evil, truth, time and good.

These reflections call into question not only Kant's moral philosophy, but any moral philosophy that reproduces the type of 'ethics' which has largely dominated European thought since the Stoics. The terms 'moral law', 'norm' or 'moral judgement' now raise suspicions of being masks of evil, even as purely conceptual forms. An equally questionable perspective is one in which 'good' and 'evil' are viewed as attributes of actions based on the idea of a freedom violently separated from nature, no longer connected to what humans *are*. Every action is an expressive movement; every action manifests something that is not absorbed in the action itself—motives, drives, reactions, interactions between the actor and their political, social and natural surroundings, but most of all the actor themselves. Humans, not actions, are good and evil. If we are to understand evil, we must pursue the question of how humans exist in nature and in society.

This is the starting point for the investigations in my anthropological texts.[12] I cannot recapitulate them in this context, and will only specify where, while following on from Kant, I part ways with him. The relative freedom of the human being is based not on the categorical imperative, but on the fact that they know they will die—that they are simultaneously distanced from the time in which they exist through the anticipation of death. Here it is fitting to use the singular 'the human being', for this is the fate of all humans. The distance from time does not place them in the timeless truth of unconditional knowledge, but it does allow them to experience

[12] See 'The Historical Nature of the Human Being', this volume, p. 0

the transience of all temporal existence directly, and not only through reflection. Through their immanent distance from time, they are distanced from their life even in the midst of it. They are not tied to their instinct mechanisms, and one could show that in humans, sensory perception and affects are also transformed in a specific way through their conscious 'being-in-time'. Thanks to their distance from time, they operate in an open manoeuvring space. This space is what gives humans their selection advantage over other animals. The manoeuvring space does not lie outside of time; it is a manoeuvring space for our temporality. It is, as we could also put it, a historical manoeuvring space. It is disclosed to us, however, by the fact that we are simultaneously distanced from the time in which we exist. We imagine that good and evil are human privileges because we experience the power of evil within this manoeuvring space. But it only gives us, as the Bible states, the possibility to recognize this power. This does not rule out the possibility that it pervades everything that is in time—in fact, it presupposes it. Time as such is negativity, and good can only appear in the medium of negativity.

In my studies on anthropology, I showed that humans—we must now speak in the plural, for they can only exist in collectives—are the living beings that are placed outside into the horizon of universal time. For them, this horizon opens up the present in which they, exposed to the interactions in the field of power, must assert themselves. In this present, they are not only at the mercy of the effects of their immediate surroundings. The past is also present for them in the implicit memory of language, without which they cannot think, in the institutions of the state and society as well as the traditions that formed them and in which they live. The consciousness of their temporality contains the ability to transform implicit into explicit memory. This reveals the countless sequences of antinomies that are ingrown in every human being. Historicity means fragmentation. It is in overcoming our inborn and imposed antinomies that the life of the individual, groups and peoples develops as 'biography'.

If humans were helplessly at the mercy of the genetic programme inscribed in them and their given living conditions with no will of their own, the development curve I call 'biography' would not come about. However, they are simultaneously distanced from the time in which they exist and the life in which they think and act through the anticipation of death. They therefore have the possibility to *shape* their biography. A manoeuvring space is opened up for them to choose, consciously or unconsciously, what they will integrate into their biography and what they will not; this manoeuvring space has a different form for every individual and in every situation. Hence we term the temporality of humans 'historical'. So the shaping of biography is the result of an uninterrupted process of integration subject to the law that once a phase has been passed through, it cannot be reversed: each phase inscribes itself into the form of the developing phenotype. This—to repeat—applies not only to individuals but also to collectives. In the field of power, the evolution of collectives exerts immense resistance against the personal shape of an individual biography. What we generally refer to with the almost absurd word 'I' contains the whole field of tension between these interdependences and the contradictions contained in them. One could describe this more precisely as the respective integration level reached by a biography. We are mentally compelled to impose on every integration an integration centre that we view as the 'subject' of the biography. It is in this sense that we then say: 'I "have" a biography.' But the 'I' that supposedly 'has' this biography is only built up through the biography, and precisely this, the establishment of the biography, always revolves around evil or good.

If we were not distanced from our life and had not been placed outside into the horizon of universal time, we could interpret the process of integration in evolutionary terms: whatever furthered self-preservation and the survival of the species would be good, and whatever was detrimental to it would be bad. But language knows what we cannot think. It associates evil with egotism. It chooses 'selflessness' as good. Humans are not only concerned with their own self-preservation; one can see this from the fact that they are willing to sacrifice their lives for another person or a cause that is sacred to them. The willingness to make sacrifices is the criterion for establishing what a person or a society truly takes seriously. Integration only succeeds in the light of the seriousness of reality. As paradoxical as it may be, this integration only occurs when people forget themselves in the service of a task, that is, when they are *not* concerned with self-preservation; if they circle their 'self', however, they fail at the task of integration. How can this paradox be explained?

Here we must recall a defining ground of human existence that we have not mentioned yet. Because humans know that they exist in time, all human feeling, thought and action is always oriented towards the future. Looking ahead to the future determines the perspectives from which humans choose what they want to integrate into their lives. As the living beings that are placed outside of themselves, they always have a wide range of possible directions in which to orient their biographies. Whoever places their life in the service of a mission has already decided for the future what they will take up into their biography and what they will discard. Those who know no task that points beyond them will drift along and become playthings of chance. They will never attain a biography that has a shape; the same applies to collectives of all kinds. One could develop a morphology of human socialization from this perspective.

The integration centre that we tend to impose on a biography as a subject, then, lies not within us but outside us in the future. For people who set themselves a great task—motherhood would be an example of such a task—this 'I' lies in a future beyond their own death. We can be mistaken in our choice of tasks. We can serve a cause that we believe to be good, but later turns out to be evil. We can fall under the spell of the masks of evil. But even then, a law emerges for each phase of our biography that transcends the evolution of other living beings: from the perspective of the chosen task, we can say for each individual phase of our biography, for every day and every hour of our life, whether we can account for them in the service of the task that has been set for us. Every state we pass through must be justified with reference to its possible integration. Thus responsibility, for both individuals and collectives, provides the criteria for ascertaining whether or not the individual or collective consciousness can vouch for its constitution.

Integration as such is not automatically good; one can set oneself evil goals. Goebbels had a biography of singular consistency and stylistic unity. Evil individuals who are integrated are not selfless but self-centred. They do not point beyond the self, which loses its meaning with death. Collectives can also have such perverted, or rather 'inverted' goals: 'Deutschland, Deutschland, über alles!', 'the dictatorship of the proletariat', 'America First'. In reality, it is always about power; it exudes such fascination that the question of the purpose for which it is used is obscured by its aura. Taken on its own, power is neither good nor evil; it is indispensable for the preservation of life. But fascination with power is always evil because it blinds us and destroys selflessness.

A morphological examination of the different manifestations of power would bring psychic mechanisms to light in both the collective and the individual consciousness with such atavistic traits that a purportedly enlightened age does not want to believe they are true. When a national football team wins a world championship, every citizen of that nation feels elevated in their sense of self. They identified with the eleven football players so deeply that they feel as if they had scored the goals themselves. When one motorist steps on the accelerator and overtakes another, they feel superior. This identification between human and machine shows that the technological world has a thoroughly magical relationship with its devices. The uncanny effect of magic in human-machine systems lies in the fact that the identification brings about a reversal of the relation between tool and will. The machine is built to serve the human being; if it is available, they operate it. They become a functionary of its system of rules; they substitute the machine for their self. This reflexively creates the illusion that the energy driving the machine is the power of the human who operates it. One finds similar identification mechanisms in the relationship of humans to capital. They are at the heart of the cult of personality and pervert the love of one's country into nationalism. Their magical character shows that they are unenlightened and unilluminated atavisms which our collective memory carries within itself.

These atavisms intersect and merge with another even more dangerous legacy of our pre-hominid ancestors. Whatever their level of intelligence, humans automatically orient themselves in the world according to the friend-foe schema that was important at earlier stages of evolution. The friend is good, the foe is evil. The current situation of the world can be described thus: the rules laid down in the friend-foe schema for distinguishing between good and evil threaten to sweep away any other knowledge about what is good and what is evil. Even Christians tend to assume automatically that a 'communist', that is, the citizen of a country with a communist government, must be evil merely for this reason. The fact that they are also a human being is no longer considered. In an increasingly polarised world, anything that is different is the enemy: other races, other nations, other religions, other economic systems, other parties. Wherever this mechanism is effective, it will disable the faculty to perceive reality as it is and to act in accordance with the facts. Every trace of humanity disappears. The friend-foe schema as such is evil, no matter where it is applied. But the power of atavisms over our consciousness reveals itself in the fact that even piety, morality or high levels of education and intelligence do not protect from its effects. There will soon be no one left who is not at their mercy.

When the friend-foe schema comes together with magical identifications, the result is war. People build nuclear bombs to reinforce the collective sense of self, using their power to annihilate their enemies if they cannot be intimidated or blackmailed.

They think they are renewing national self-confidence and national society by ruining themselves economically with the excess production of weapons of mass destruction. Here all the demons of evil are unleashed. It is futile to attempt making people aware that both magical identifications and those dictated by the friend-foe schema are based on fictitious concepts of the 'self' and its 'identity'. With recourse to the section on metaphysics, we can say that identity is the signature of evil. It is the false that appears in the guise of the truth.

Identity is negation of time. Identity is negation of death. Whoever negates time and death no longer perceives reality. Transcendental consciousness in its timeless identity with itself can neither love nor suffer. Hope is unknown to it; but where hope dies, evil reigns. In Dante, the gate to hell bears the inscription 'Lasciate ogni speranza, voi ch'entrate'—'Abandon hope all ye who enter here'. The mythical image of the inferno depicts the nature of the realm of evil in the midst of our world: where evil takes power, hope falls silent. We let hope evaporate in our illusions; it suffocates in the emotions of the collective consciousness. It awakens and grows strong when we soberly contemplate the seriousness of reality and the wealth of its antinomies. Here reality reveals itself to us in the light of death, and simultaneously points to a future for which we bear responsibility even if we will no longer be there to experience it ourselves. We only act on this responsibility, we only live up to it, if we are carried by the force of a hope that no longer concerns our own existence. Hope opens the heart and the spirit to good; evil is the power that opposes hope.

On Georg Picht

Georg Picht was born on 9 July 1913 in the French city of Strasbourg, which belonged to the German Empire at the time. His father Werner Picht, a jurist and sociologist, served as an officer at the front during the First World War before becoming head of adult education at the Prussian Ministry of Culture; he later moved to Paris, where he headed the university department at the International Institute of Intellectual Cooperation, and left behind a literary and essayistic output dealing with the themes of soldierly life and Christian teachings, but also such figures as Stefan George and Albert Schweitzer. On his mother's side, Georg Picht was descended from a family of scholars: his great-grandfather Ernst Curtius was a classical scholar and directed the German excavations in Olympia; his grandfather Friedrich Curtius, an administrative jurist and later President of the Supreme Consistory of the Protestant Church of the Augsburg Confession of Alsace and Lorraine, authored texts on political matters; his uncle Ernst-Robert Curtius was a leading scholar of Romance languages.

The members of the Curtius household had vibrant intellectual exchanges and friendships with such figures as Franz Rosenzweig, Hans and Rudolf Ehrenberg, Eugen Rosenstock-Huessy, Albert Schweitzer, Charles du Bos, Hermann Kantorowicz, Viktor von Weizsäcker and others. There were also connections to Stefan George's circle.

Georg Picht, who suffered from delicate health, spent the majority of his childhood from the age of seven in the Black Forest village of Hinterzarten, which remained his primary place of residence throughout his life. For several years, he received his secondary schooling at home from the classicist and numismatist Josef Liegle. He abandoned his original plan to become a pianist after meeting his later wife, the

E. Rudolph and J. Picht (eds.), *Georg Picht: A Pioneer in Philosophy, Politics and the Arts*, Pioneers in Arts, Humanities, Science, Engineering, Practice 19,
https://doi.org/10.1007/978-3-030-31790-4

pianist and harpsichordist Edith Axenfeld. He studied classics and philosophy in Freiburg, Kiel and Berlin, and his teachers included Wolfgang Schadewaldt, Eduard Fraenkel, Martin Heidegger und Wilhelm Szilasi. After working briefly in the editorial committee for the Church Fathers edition at the Berlin Academy of Sciences, he taught ancient languages from 1939 at Birklehof School in Hinterzarten, which was a private boarding school in the reform education tradition until the Nazis took it over in 1942. The same year, he gained his doctorate with a thesis entitled 'Die Grundlagen der Ethik des Panaitios' [The Foundations of the Ethics of Panaetius]. Deemed unfit for military service on account of his health, Picht was an assistant at the University of Freiburg's classics department until the end of the war. Profoundly disillusioned by the inability of the universities to resist the Nazi dictatorship and the expulsion of Jewish colleagues, Picht did not initially intend to pursue an academic career after the war and instead became head teacher at Birklehof. There he tried to realize his classically-based ideal of responsible humanism as an educator. From the early 1950s on, he also headed the 'Plato Archive', a project to compile a Plato lexicon that was financed by the German Research Foundation but remained unfinished. From 1953 he was a member of the German Education Committee, which had been set up by Theodor Heuss, President of the Federal Republic of Germany. From 1958 until his death he was director of the newly-founded Protestant Institute for Interdisciplinary Research, whose declared aims include 'helping the church in its engagement with the world'. Picht appointed not only theologians and philosophers but also physicists, biologists, jurists, political scientists and economists as staff members. Under his directorship, the centre produced acclaimed studies on the socio-political situation of the army, the problem of nuclear weapons, environmental matters and other issues. Picht established 'peace research' as a project area and turned the research centre into an institute for political advice. In 1961 he was a signatory of the 'Tübingen Memorandum', which received considerable attention and argued for the recognition of the Oder-Neisse border with Poland. With his series of articles entitled 'Die deutsche Bildungskatastrophe' [The German Education Disaster] (1964), in which he demanded that academic education be made more widely available, he set off controversies that have continued to this day. This allowed him to exert considerable influence on the discussion on education policy in Germany. In 1965 he was awarded the Theodor Heuss Prize for this courageous initiative. The lecture cycle 'Mut zur Utopie' [Courage for Utopia] was translated into several languages. Through his lifelong friendship with the physicist and philosopher Carl-Friedrich von Weizsäcker, Picht was confronted early on with developments in modern natural science and technology as well as their destructive potential. The question of the conditions of possibility for human responsibility in history, which he posed against this background, is one of the central motifs in his thought.

In 1965 Picht accepted the newly-established professorship for the philosophy of religion in the theology department at the University of Heidelberg, a position he held until 1978. In the lectures he gave there, he was able to expand at length on the content that was sometimes presented in condensed form in his numerous essays (of which this volume contains a selection). Some of them deal with the work and reception history of individual philosophers (Plato, Aristotle, Kant, Nietzsche), while others address topics that arose from the focuses of his research in different problem areas. These topics included knowledge of the future, art and myth, the history of the concept of nature or the philosophy of history. His thinking is always characterized by an inseparable unity of historical and systematic perspectives. Picht attempted to develop his own philosophy of time in his final book, entitled *Von der Zeit* [On Time], but the attempt remained incomplete: he died in Hinterzarten on 11 August 1982, not long after his 69th birthday.

Selected Bibliography

1. Essay collections

1.1 *Die Verantwortung des Geistes* [The Responsibility of the Spirit] (1965)

1.2 *Wahrheit – Vernunft – Verantwortung* [Truth – Reason – Responsibility] (1969)

1.3 *Mut zur Utopie* [Courage for Utopia] (1969)

1.4 *Hier und Jetzt: Philosophieren nach Auschwitz und Hiroshima* [Here and Now: Philosophizing After Auschwitz and Hiroshima] (1980/81, 2 vols.)

2. Lectures and Writings (ed. Constanze Eisenbart)

2.1 *Kants Religionsphilosophie* [Kant's Philosophy of Religion] (1985)

2.2 *Kunst und Mythos* [Art and Myth] (1986)

2.3 *Aristoteles' De Anima* [Aristoteles' De Anima] (1987)

2.4 *Nietzsche* (1988)

2.5 *Der Begriff der Natur und seine Geschichte* [The Concept of Nature and Its History] (1989)

2.6 *Platons Dialoge Nomoi und Symposion* [Plato's Dialogues *Laws* and *The Symposium*] (1990)

2.7 *Glauben und Wissen* [Faith and Knowledge] (1991)

2.8 *Zukunft und Utopie* [Future and Utopia] (1992)

2.9 *Geschichte und Gegenwart* [History and the Present] (1993)

2.10 *Die Fundamente der griechischen Ontologie* [The Foundations of Greek Ontology] (1995)

2.11 *Von der Zeit* [On Time] (1999)

E. Rudolph and J. Picht (eds.), *Georg Picht: A Pioneer in Philosophy, Politics and the Arts*, Pioneers in Arts, Humanities, Science, Engineering, Practice 19,
https://doi.org/10.1007/978-3-030-31790-4

3. Other Writings

3.1 *Die Grundlagen der Ethik des Panaitios* [The Foundations of the Ethics of Panaetius] (unpublished PhD thesis)

3.2 *Kommentar zu Platons Dialog Laches* [Commentary on Plato's Dialogue *Laches*] (unpublished)

3.3 *Naturwissenschaft und Bildung* [Natural Science and Education] (1954)

3.4 *Unterwegs zu neuen Leitbildern* [Towards New Models] (1957)

3.5 *Die deutsche Bildungskatastrophe* [The German Education Disaster] (1964)

3.6 *Der Gott der Philosophen und die Wissenschaft der Neuzeit* [The God of the Philosophers and the Science of the Modern Age] (1966)

3.7 *Theologie – was ist das?* [Theology – What Is It?] (1977)

3.8 *Ist Humanökologie möglich?* [Is Human Ecology Possible?] (1979)

3.9 *Zum Begriff des Maßes* [On the Concept of Measure] (1979)

3.10 *Utopie und Hoffnung* [Utopia and Hope] (1979)

On the Editors

Enno Rudolph (born 1945) studied philosophy and Protestant theology at the Universities of Münster and Heidelberg. Diploma (1970), PhD on Kant (1974), Habilitation on Aristotle (1983) both at the University of Heidelberg. He taught as an extraordinary Assoc. Professor at the University of Heidelberg and was head of philosophical research at FESt *(Protestant Institute for interdisciplinary Research)* partly together with Georg Picht and Carl Friedrich von Weizsäcker. He had several guest professorships (Rio de Janeiro, Florence, Berlin, Jena and Tübingen) and research fellowships (Paris, Wien, Pisa, Torino). Since 1999 he has been Full Professor for philosophy of politics and philosophy of culture at the University of Lucerne (Switzerland) where he is now Emeritus.

His major publications are: *Odyssée des Individuums* [Odyssée of the Individual] (1991), *Ernst Cassirer im Kontext* [Ernst Cassirer in Context] (2003), *Theologie - diesseits des Dogmas* [Theology beyond dogmas] (2006), *Wege der Macht* [Paths to power] (2017), *Der Europäer Erasmus von Rotterdam* [The European Erasmus of Rotterdam] (2019), *Machiavelli* (2021).

Address: Prof. Dr. Enno Rudolph, Adalbert-Stifter-Strasse 17, 69151 Neckargemünd, Germany

Email: <enno.rudolph@unilu.ch>

Website: CV and comprehensive list of publications: <https://www.unilu.ch/fileadmin/fakultaeten/ksf/institute/kuwifo/dok/CV_Publist_Rudolph_Word.pdf>.

E. Rudolph and J. Picht (eds.), *Georg Picht: A Pioneer in Philosophy, Politics and the Arts*, Pioneers in Arts, Humanities, Science, Engineering, Practice 19,
https://doi.org/10.1007/978-3-030-31790-4

Johannes Picht (born 1954) was first trained as a pianist before he turned to philosophy and then medicine, studying in Berlin, Munich and Freiburg/Breisgau. He specialized in Internal Medicine and in Psychosomatic Medicine and became a psychoanalyst, in which profession he has now worked in private practice for near to 25 years. He is a member of the German and the International Psychoanalytic Association and appointed supervisor and training analyst. He is also editor of leading psychoanalytic journals such as "Jahrbuch der Psychoanalyse" (2013-2017) and "PSYCHE – Zeitschrift für Psychoanalyse und ihre Anwendungen" (since 2017 as editor-in-chief). He has edited two volumes on music and psychoanalysis and published numerous articles on music, psychosomatics, and psychoanalytic theory and technique.

Major titles: "Beethoven und die Krise des Subjekts" [Beethoven and the crisis of the subject] (2007/08); "Zur ethischen Grundlegung der Abstinenz" [On the ethical foundation of abstinence] (2014); "Sprache, Musik und das Unbewusste" [Language, Music, and the Unconscious] (2015); "Dimensionen des Geschehens und das Phantasma der Begegnung" [Dimensions of proceedings and the phantasma of encounter] (2018); "Das Private und seine Erforschung" [Privacy and its investigation] (2020). For a full list of publication see <http://www.johannes-picht.de/publikationen/>.

Address: Dr. Johannes Picht, Bellinger Str. 1, 79418 Schliengen, Germany
e-mail: <johannes.picht@web.de>
website: <http://www.johannes-picht.de>.

Author Index

A
Abbé de Saint-Pierre, 91
Adorno, Theodor W., 1–4, 94, 178, 179, 181, 183–186
Aeschylus, 23, 204
Agamemnon, 193, 194
Alcibiades, 43, 44, 47, 48, 50–53, 56
Aletheia, 25, 34, 35, 55
Alexander the Great, 204
Ananke, 22–24, 29, 39
Anaximander, 61
Anaxogoras, 13
Antigone, 116, 118
Apollo, 13, 204
Ares, 203–205
Aristotle, 8, 9, 18, 23, 45–48, 52, 57–72, 108, 110, 125, 126, 132–139, 142, 144, 145, 152, 161, 162, 167, 168, 179, 208, 217, 231
Ashton, E. B., 4
Athena, 12, 203, 206, 210
Augustine, 109, 112, 114

B
Bacon, Francis, 132
Balzac, Honoré de, 146
Baumgarten, Alexander Gottlieb, 201
Becker, Albrecht, 63
Becker, Carl, 207
Beethoven, Ludwig van, 207
Benedict, Ruth, 107, 108, 164
Boethius, 55
Boll, Franz, 28
Bollnow, Otto F., 163
Boulez, Pierre, 3
Bowring, Edgar A., 193
Brauch, Hans Günter, 215
Brooks, Charles T., 199

C
Calchas, 13
Charmides, 52, 54, 55, 134
Cicero, Marcus Tullius, 108, 109, 111
Cohler, Anne M., 108
Comte, Auguste, 165
Creon, 116
Crisp, Roger, 45
Critias, 28, 29
Cusa, Nicholas of, 153

D
Dante, Alighieri, 55, 227
Darwin, Charles, 163, 164, 167, 215
Deimos, 203, 204
Demokrit, 11, 46
Descartes, René, 163, 164, 168, 222
Desplant, Michel, 220
Diels, Hermann Alexander, 15, 17, 18, 58
Diggle, James, 47
di Giovanni, George, 221
Dike, 21, 22, 39
Dilthey, Wilhelm, 110, 164
Dionysus, 44
Duchrow, Ulrich, 109

E. Rudolph and J. Picht (eds.), *Georg Picht: A Pioneer in Philosophy, Politics and the Arts*, Pioneers in Arts, Humanities, Science, Engineering, Practice 19,
https://doi.org/10.1007/978-3-030-31790-4

E

Eckhart, Meister, 57
Empedocles, 13, 134
Epimetheus, 136
Eris, 203–207

F

Faust, 211, 212
Ferrari, G. R. F., 47
Fichte, Johann Gottlieb, 50, 76
Forster, Eduard, 66
Fränkel, Hermann, 8, 16, 24, 39, 41
Freud, Sigmund, 164, 215
Friesenhahn, Ernst, 114
Fritz, Kurt von, 11

G

Galileo di Galilei, 132
Gehlen, Arnold, 111, 136, 163, 164
Goethe, Johann Wolfgang von, 163, 180, 210–213
Gregor, Mary, 80, 85, 130, 140
Griffith, Tom, 47
Guyer, Paul, 77, 79, 130, 153

H

Hartmann, Nicolai, 146
Hatfield, Gary, 101
Hazard, Paul, 76, 110
Hegel, Georg Wilhelm Friedrich, 2, 3, 9, 49, 51, 76, 87, 90, 125, 126, 132, 133, 137, 139, 147, 151, 162, 168, 169, 179, 181, 185, 189, 190, 195, 199–201, 220, 221
Heidegger, Martin, 46, 94, 169–171, 221, 230
Heracles, 204
Heraclitus/ Heraclites, 13, 34, 134
Herder, Johann Gottfried, 76, 77, 100, 136, 163
Hermes, 26, 136
Herz, Marcus, 94
Hesiod, 8, 10, 12–17, 20–22, 27, 30, 32, 137, 152, 206
Hobbes, Thomas, 163
Holstein-Augustenburg, Duke Christian von, 189
Homer, 8–11, 13–16, 18, 20, 21, 27, 30, 32, 33, 36, 37, 42, 152, 203–207, 209, 210
Hong, Edna H., 50
Hong, Howard V., 50
Horace, 194
Horkheimer, Max, 183
Howatson, M. C., 43
Hullot-Kentor, Robert, 178

I

Illies, Joachim, 161

J

Jaeger, Werner, 32, 35
Jean Paul, 199–202
Jellinek, Georg, 111, 112
Jesus of Nazareth, 173

K

Kant, Immanuel, 3, 8, 50, 75–105, 110, 115, 118, 126, 130–133, 139–145, 153–156, 158, 159, 163, 164, 168, 169, 180, 185, 186, 193, 208, 217, 220–223, 231
Kaser, Max, 108, 111
Kenny, Anthony, 208
Kierkegaard, Søren, 43, 49, 50, 56, 217
Kline, A. S., 212
Knox, Thomas Malcolm, 51, 189, 195, 200
Kranz, Walther, 38, 39
Kriele, Martin, 111
Kullmann, Wolfgang, 26, 33

L

Laches, 52
Leakey, Richard, 165
Leibniz, Gottfried Wilhelm von, 163, 170, 171, 178, 179, 219, 220
Livingstone, Rodney, 4
Lombardo, Stanley, 204
Lorenz, Konrad, 164, 216
Louden, Robert B., 88
Luther, Wilhelm, 11, 46, 220

M

Mallarmé, Stéphane, 200
Marx, Karl, 3, 76, 94, 163, 164, 178, 185, 190, 215
Matthews, Eric, 79
McMillan, Clare, 163
Meineke, Johann Albrecht Friedrich August, 35

Menzel, Eberhard, 107, 118, 119
Miller, Arnold V., 9
Miller, Basia Carolyn, 108
Mnemosyne, 10, 11, 21
Moira, 20, 22, 39
Montesquieu Baron de, 108, 114, 115
Moore, George Edward, 146
Muse, 10–15, 17, 18, 20–22, 27, 30, 32, 33, 42, 55, 194, 210

N
Napoleon Bonaparte, 192
Newton, Isaac, 76, 212
Nietzsche, Friedrich, 75, 76, 94, 125, 136, 146, 147, 164, 215, 217, 220, 231
Nilsson, Martin P., 21, 35
Nisbet, H. Barry, 78

O
Odysseus, 12, 28
Orestes, 194, 195

P
Panaetius/Panaitios, 111, 230
Pandora, 210
Parmenides, 7, 8, 10, 13, 14, 16–38, 40, 41, 55, 134, 151, 152
Perry, Ralph Barton, 146
Pfau, Thomas, 222
Phaedrus, 23
Phemios, 11
Philebus, 214
Phobos, 203–209, 214
Phoebus, 213
Picht, Johannes, 1, 7, 43, 57, 75, 107, 123, 129, 151, 161, 177, 189, 197, 203
Pickford, Henry W., 1
Pillemer, Karl, 163
Pindar, 21, 34, 35, 172
Plato, 8–11, 13, 15, 18, 23, 24, 28, 43, 47, 49, 51–53, 55, 56, 63–65, 71, 91, 98, 125, 133–140, 142–144, 146, 162, 164, 167, 168, 172, 179, 203, 208, 210, 214, 217, 230, 231
Plessner, Helmuth, 163, 169
Plutarch, 204
Polykleitus, 144
Portmann, Adolf, 136
Prometheus, 23, 136, 210
Protagoras, 54, 134–137, 142
Pythagoras, 61

R
Raiser, Ludwig, 117
Ranke, Franz Leopold von, 112
Reinhardt, Karl, 207
Reiss, Hans S., 78
Ribet, 146
Roček, Roman, 163
Ross, W.D., 58, 69
Rousseau, Jean-Jacques, 91, 190, 198
Rudolph, Enno, 1, 7, 43, 57, 75, 107, 123, 129, 151, 161, 177, 189, 197, 203

S
Schadewaldt, Wolfgang, 208, 230
Schatz, Oskar, 163
Scheler, Max, 96, 163, 169
Schelling, Friedrich Wilhelm Joseph, 57, 76, 180, 201, 217, 221, 222
Schiller, Friedrich, 119, 189–195, 201
Schlegel, Friedrich, 49–51
Schmitt, Carl, 112, 113
Seitzer, Jeffrey, 112
Shaftesbury, Anthony, 88
Sheffield, Frisbee C. C., 43
Sibree, John, 220
Snell, Bruno, 189
Sokrates, 16, 43
Solger, Karl Wilhelm Ferdinand, 49, 50
Sprague, Rosamund Kent, 52
Stone, Harold Samuel, 108

T
Tarán, Leonardo, 35
Theaetetus, 135
Themistius, 65
Theophrastes/ Theophrastus, 47
Thrasymachus, 47, 48
Timaeus, 9, 71
Torricelli, Evangelista, 132
Tribe, Keith, 119

V
Voltaire, François-Marie Arouet, 220

W
West, Martin Litchfield, 206
Wieser, Wolfgang, 171
Wood, Allen W., 77, 130, 153, 221
Wormeley,Katharine Prescott, 146

X
Xenocrates, 133
Xenophanes, 8, 9, 13–23, 27, 30, 33, 36, 48

Y
Young, J. Michael, 77, 141

Z
Zenon, 26, 33
Zeus, 10, 12, 13, 16, 17, 30, 32, 34, 35, 136, 204
Zöller, Günter, 88

Subject Index

A
Absolute, absolutism, 3, 4, 7, 50, 51, 94, 110, 116–118, 126, 142, 146, 152, 155, 165, 178, 179, 183, 185, 189, 192–196, 199–202, 217, 221, 222
Aesthetics, 2, 3, 51, 119, 146, 178, 179, 183, 184, 187, 189, 190, 193, 194, 197, 200–202, 208, 209, 214
Alienation, 181, 190, 191, 198, 200
Antiquity, 198
Appearance, 7, 9, 30–33, 35, 39, 42, 43, 55, 56, 78–81, 94, 99, 103, 129, 135, 143, 151–155, 157, 159, 193–195, 200, 201, 221
A priori, 9, 11, 81, 87, 92, 100–105, 131, 132, 155, 166, 169, 185, 186, 223
Aretē, 45, 54, 137
Art, 3, 11, 15, 16, 47, 48, 51, 133, 136, 138, 177–180, 182–184, 186–190, 192–203, 210, 212, 213, 217, 218
Autonomy, autonomous, 1, 94, 95, 116, 117, 141, 143, 168, 169, 177, 179, 182, 183, 195, 222

B
Beauty, 44, 48, 49, 51, 189, 192, 193, 195, 200, 201, 207, 215
Being, 94, 169
Bios, 138, 166, 171

C
Capitalism, 126, 146, 186, 192
Causality, 94, 96, 101, 155, 157
Christian, Christendom, 45, 46, 76, 107, 113, 114, 118, 119, 129, 163, 168, 177, 215, 219, 226, 229
Concept, conceptualization, 2, 3, 8, 26, 29, 36, 45, 48–51, 57–59, 61–63, 65, 67–72, 76, 77, 79, 81, 82, 84, 85, 88, 93–95, 97, 100, 110, 111, 114, 115, 118, 120, 121, 126–132, 136, 138, 139, 141, 142, 144–147, 152–155, 158–160, 164, 169, 179, 182, 185–187, 189, 194, 195, 202, 206, 207, 216, 217, 219, 222, 227, 231
Conscience, 46, 76, 111–114, 120, 198, 201, 215, 219
Consciousness, 2, 3, 45, 49, 50, 76, 97, 102, 107, 108, 115, 119, 123, 124, 126–128, 137, 162, 163, 167, 168, 177–180, 183–188, 197, 198, 201, 202, 208, 215, 216, 218, 223–227
Constitution, 58, 71, 87, 89, 93, 96–99, 107, 112, 113, 115, 117, 120, 131, 185, 186, 192, 200, 225
Continuum, continuity, 26, 27, 31, 61, 62, 64, 108, 125, 139, 140, 143, 151, 152, 155
Contradiction, principle of, 217
Cosmos, 15, 71, 109, 110, 131, 151, 153, 172, 219
Critical theory, 183
Critique, 1, 3, 8, 77–79, 81–87, 92–94, 96–98, 101–104, 126, 128, 130–132, 140, 142, 143, 145, 153, 163, 178–180, 183, 184, 220, 222
Culture, cultural, 46, 76, 87, 110, 115, 120, 124, 125, 130, 135, 138, 147, 161,

E. Rudolph and J. Picht (eds.), *Georg Picht: A Pioneer in Philosophy, Politics and the Arts*, Pioneers in Arts, Humanities, Science, Engineering, Practice 19,
https://doi.org/10.1007/978-3-030-31790-4

164, 166, 169, 171, 187, 190, 198, 215–217

D
Determination, determinism, 3, 55, 60–62, 64, 97, 147, 153–157, 159, 169, 208, 221, 223
Dialectic, dialectics, 2–4, 78, 90, 94, 126, 154, 179, 185, 186, 214

E
Ecological, ecologics, 120, 121, 146, 147, 178, 180, 181, 219
Economy, economics, 2, 76, 113–115, 121, 128–131, 136–142, 146, 147, 166, 169, 177, 180, 181, 187, 191, 216, 226
Empirical, empiricism, 3, 51, 79–82, 84, 86, 97, 99, 101, 110, 126, 130, 131, 154, 164–167, 170, 183, 185, 186, 223
Energeia, 23, 57–59, 62, 63, 67–72
Enlightenment, 77, 93–95, 123, 124, 165, 190, 192, 220
Entelecheia, 58, 59, 62, 63, 65–67, 69, 71, 72
Environment, 171, 172
Eternity, 13, 18, 27, 42, 151, 152, 154, 171
Ethics, 45, 47–49, 52, 60, 88, 96, 98, 107, 109, 111, 115, 117, 118, 125, 129–147, 166, 168, 223
Evolution, 100, 164–167, 171, 172, 182, 217–219, 224–226
Existence, existent, 13, 23, 24, 27–29, 31, 34, 40, 41, 44, 50, 51, 56, 57, 60–68, 70, 79, 81, 82, 84, 86, 87, 89, 91–95, 98, 99, 102–104, 118, 131, 132, 135, 137, 139, 141–144, 147, 153–156, 158, 169, 173, 178, 180, 183–185, 190, 193–195, 198, 209, 210, 212, 218–220, 224, 225, 227

F
Faith, 94, 110, 116, 118, 123, 165, 177, 197, 198, 220
Freedom, 1, 3, 4, 8, 51, 53, 76, 78–100, 103, 104, 110–114, 117, 119, 120, 124, 128, 130–132, 140–143, 145, 155, 177, 183, 186, 189–195, 200–202, 219–223
Future, 13, 18, 27, 28, 30–32, 42, 78, 100, 101, 104, 105, 113, 147, 151, 154, 155, 157–160, 170–172, 182, 214, 225, 227

G
Genetic, genetics, 218, 224
God, 11, 13–19, 21, 27, 28, 31, 33, 42, 44–47, 50, 54–57, 85, 88, 94, 104, 109, 116, 118, 124, 126, 133–137, 139, 142, 151–154, 160, 186, 198–210, 213, 219–222
Good, 17, 43, 48, 50, 51, 53, 55, 85, 88, 113, 115, 133–135, 137–139, 145–147, 153, 180, 181, 215, 217–221, 223–227

H
Harmony, 52, 88, 109, 114, 178, 179, 181, 191–194, 201, 219
History, historical, 1, 2, 7–10, 45, 54, 55, 57, 61, 75–80, 82–88, 90–96, 100–105, 107–110, 112–114, 117, 118, 125, 127–129, 131, 134, 136, 140, 143, 145, 146, 155, 159, 161–164, 168–173, 181, 182, 184, 186, 187, 191, 193–195, 197, 202, 203, 207, 215, 216, 220, 221, 224, 230, 231
Horizon, 3, 8, 13, 46, 49, 57, 59, 72, 75–77, 100, 123, 127, 128, 133, 147, 151, 152, 154, 158–160, 166, 168, 182, 209, 221, 224, 225
Human, humanity, 80, 100, 104, 112, 115–121, 136, 147, 190, 191, 197, 198, 206, 216, 226

I
Idea, 1, 4, 26, 50, 51, 68, 71, 76–80, 82–85, 87–90, 93, 96, 98, 100, 102–104, 109, 112–115, 117, 126, 132, 133, 138, 139, 142, 143, 145, 146, 152, 153, 160, 163, 164, 167–169, 178–180, 184–186, 189, 193–195, 198, 199, 201, 202, 205, 207, 212, 213, 219, 220, 222, 223
Identity, 3, 19, 65, 131, 141, 142, 144, 151, 152, 154, 159, 198, 200, 222, 223, 227
Imperative, categorical/hypothetical, 81, 96–99, 110, 131, 141–145, 193, 223
Individual, individuality, 171, 189, 198, 201, 218, 223

J
Judgment, 79, 83, 85–87

K
Knowledge, 2, 3, 8–11, 14–18, 20–22, 30, 31, 33, 47, 48, 54–56, 86, 90, 91, 94, 95, 97, 100, 102–104, 110, 117, 118, 123, 126, 130, 131, 133–136, 138–143, 152, 157, 162, 163, 165, 166, 168–171, 173, 178, 180, 185, 195, 217, 219, 221–223, 226

L
Language, 4, 21, 23, 24, 32, 46, 54, 56, 57, 60, 63, 68, 71, 90, 99, 117, 130, 146, 167, 168, 186, 197, 198, 203, 204, 206, 207, 213, 224, 225
Law, 50, 51, 75, 78–85, 87, 89, 90, 95–102, 104, 108–118, 126, 130–132, 137, 139–141, 143, 146, 151, 155, 158, 163, 164, 167–169, 178, 180, 181, 183, 187, 190–192, 194, 195, 217, 219, 223–225
Liberal, liberalism, 112–114, 180, 186, 192, 196
Literature, 15, 144, 180, 184, 198–200, 203, 217
Logic, logos, 8–10, 32, 36, 38–40, 42, 60, 61, 64, 65, 109, 111, 114, 134, 139, 144, 147, 161, 162, 164, 166–169

M
Material, materialism, 61, 78, 92, 96, 107, 119, 130, 135, 137, 141, 156, 158, 185, 186, 194
Mathematics, 81, 125, 132, 134, 165, 168, 170, 171, 222
Measure, 15, 54, 59, 134–136, 138, 139, 141, 142, 180, 182, 207
Memory, 10–12, 31, 114, 127, 128, 163, 172, 186, 191, 198, 218, 220, 224, 226
Metaphysics, 18, 58, 59, 62–64, 67, 70–72, 80, 81, 86, 89, 95–98, 101, 102, 130–132, 138, 140, 142, 145
Metaphysics, 3, 4, 18, 55, 58, 59, 62–64, 67, 70–72, 75, 76, 78, 80, 81, 85, 86, 89, 94–98, 101, 102, 104, 111, 118–120, 124–127, 129–133, 138, 140, 142–147, 151–153, 159, 160, 162, 163, 165, 166, 181, 185, 186, 195, 217, 220, 221, 227
Method, methodology, 2, 3, 8, 44, 57, 78, 82, 89, 92, 107, 120, 121, 124, 132, 134, 139, 145, 162, 164, 165, 172, 181, 182
Modernity, modern times, 194
Moral, morality, 45, 46, 51, 77, 79–81, 85, 86, 89, 94–97, 102, 107, 110, 113, 114, 116, 119, 120, 130–132, 135, 137, 140–143, 145, 146, 164, 190–193, 195, 215, 219, 222, 223, 226
Mortal, mortality, 9, 19–22, 27, 28, 33, 35, 37–39, 54–56, 116, 135, 136, 139
Music, 3, 4, 198–200, 208, 213, 217
Myth, 10, 123, 136, 172, 177, 187, 188, 203, 214, 218

N
Nature, 13, 17, 21, 22, 26, 28–30, 37, 39, 40, 44, 46–49, 51, 53, 54, 57–63, 65, 67, 68, 70, 78–92, 96, 98, 100, 103, 105, 108–119, 124, 126, 127, 130–132, 134–137, 140–147, 151–158, 161–170, 172, 173, 177, 178, 180–184, 186, 187, 190–195, 197–201, 208, 210, 212–214, 217–219, 221–223, 227
Necessary, necessity, 9, 15, 20, 22, 23, 27–29, 36, 38, 39, 41, 49, 62, 66, 76, 78, 79, 81, 83, 85, 87, 89, 90, 98, 108, 113, 116, 124, 130, 131, 137, 138, 140–142, 156–159, 164, 170, 180, 185, 191, 195, 201, 219
Negative, negativity, negation, 2–4, 7, 50, 51, 61, 63, 64, 94, 152, 154, 169, 171, 172, 179, 185, 186, 195, 200, 219, 224, 227
Nihilism, 147
Nominalism, 153
Norm, normative, 48, 54, 79, 98, 107, 110, 115–118, 136, 141, 143–145, 223
Nous, 8, 11, 14, 16–20, 28–37, 42, 71, 72, 133

O
Object, 1, 2, 39, 48, 51, 56, 61, 63, 81, 84, 96, 130, 133, 135, 140, 141, 151, 153, 155, 161–164, 166, 168, 170, 172, 181, 201, 206, 209, 214, 216

Ontology, 8, 10, 26, 60, 70, 94, 130, 152, 156, 169

P

Past, 18, 28, 30, 31, 42, 91, 100, 119, 151, 154, 155, 157, 158, 169, 170, 178, 182, 186, 187, 211, 224
Phenomena, phenomenality, 10, 78, 123, 140, 152, 155, 158–160, 167, 168, 170, 184, 192, 198, 203, 206
Philosophie, philosophical, 4, 55, 57, 75–78, 80, 82, 85, 87, 88, 90, 93, 94, 104, 107, 109, 111, 113, 115, 117, 119, 123–126, 129–131, 135, 139, 145, 147, 152, 153, 161, 163, 178, 182, 214, 217, 221
Physics, 58, 59, 61, 62, 64, 65, 69–71, 80, 89, 114, 124, 125, 127, 130–136, 139–141, 145, 147, 153–156, 158, 159, 170, 171, 182, 183, 191
Physis, 11, 14–16, 32, 53, 59, 60, 64, 65, 70, 71, 109, 133–135, 139, 141
Poiesis, 179
Polis, 136–138, 161
Possible, possibility, 1, 7, 10, 17, 18, 24, 25, 31, 39, 41, 50, 52, 55, 62, 70, 71, 76–79, 81–83, 85, 86, 91, 92, 96–100, 102, 110, 114, 118, 131, 133–135, 141, 152, 154, 156–160, 168, 170, 185–187, 190, 208, 217–219, 221, 222, 224, 230
Power, 1, 3, 33, 34, 39, 48, 50, 53, 55, 76–79, 83, 85–87, 89, 95, 99, 102, 116, 123–128, 136, 138, 143, 145–147, 152, 153, 163, 169, 177, 180, 183, 184, 190, 192, 195–197, 203–210, 212, 213, 216, 217, 222, 224, 226, 227
Praxis, 66, 67, 76, 110, 120, 126, 127, 168, 193
Process, 3, 16, 62, 66, 79, 80, 82–87, 89–91, 93, 94, 99, 101–104, 108, 114, 115, 123, 127, 155, 158, 162–164, 169, 172, 178–182, 187, 191, 201, 206–208, 210, 218, 220, 224, 225
Production, 66, 87, 128, 135, 137, 178–184, 186–188, 190, 191, 227
Progress, 80, 86, 91, 94, 100–104, 208, 220
Project, projection, 104, 126, 164, 167–169, 185, 187, 193, 230
Psyche, 67, 72, 162
Psychoanalysis, 3, 164, 172, 214

R

Rational, rationality, 79, 85, 94, 96–99, 103, 112, 115, 117, 130, 140, 141, 163, 165, 167–169, 192, 206, 207
Reality, 1, 49, 51, 53, 57, 67, 68, 85, 115, 123, 124, 128, 135, 138, 140, 143, 145, 167, 170, 183, 185, 194, 205–207, 209, 210, 216, 217, 219, 221, 225–227
Reason, 2, 7, 8, 17, 31, 45, 52, 56, 61, 66, 71, 75, 77–82, 84–105, 108–111, 113, 115–119, 124, 125, 128, 130–132, 135, 140–145, 151–155, 157, 159, 160, 165, 167–169, 189, 192–194, 214, 218, 220–223, 226
Reflection, 3, 43, 46, 58, 76, 77, 83, 86, 87, 107, 119, 121, 129, 139, 147, 154, 168, 169, 178, 179, 183–186, 200, 201, 208, 212, 216, 223, 224
Regulative, regulatory, 84, 96, 145, 180
Religion, religous, 21, 35, 46, 76, 93, 94, 109–112, 115, 117, 120, 130, 135, 139, 164, 177, 195, 198, 216, 220, 221, 226, 230
Representation, 51, 96, 97, 108, 114, 140, 154, 155, 179, 191, 193, 196–200, 208, 217, 222
Res cogitans/extensa, 163, 168, 172
Responsibility, 225, 227
Revolution, 93–95, 100, 101, 103, 115, 117, 125, 132, 140, 178, 189–194, 197–199, 201
Right, human rights, 107–121

S

Sceptical, scepticism, 54, 101, 119, 135, 163
Science, 3, 7–9, 55, 59–61, 68, 76, 77, 80, 101, 118, 125–128, 130–139, 141–143, 145, 147, 152, 155, 156, 161, 162, 164–170, 177, 178, 180, 183, 191, 192, 210, 216–218
Socialism, 126, 186
Society, 1–4, 47–49, 79, 80, 82, 87–90, 98–100, 103, 113–115, 117, 123, 124, 126–128, 130, 140, 161, 162, 170, 177–187, 190–196, 215, 216, 223–225, 227
Sophist, sophism, 18, 48, 54, 134, 136
Soul, cf psyche, 162
Space, spatial, 8, 11, 21, 23, 26, 27, 58, 61, 78, 85, 109, 114, 115, 134, 151, 152,

157–159, 170, 195, 201, 211, 215, 224
Spirit, 1–4, 9, 13, 89, 90, 102, 108, 110, 116, 117, 124, 133, 140, 155, 163, 166, 167, 178, 185, 192, 195, 198, 200, 202, 213, 217, 220–222, 227
State, 15, 18, 19, 23, 30, 33, 40, 41, 46, 49, 50, 54–56, 58, 60–64, 68, 76, 82–84, 89, 90, 92, 93, 98–100, 104, 108, 110–118, 120, 121, 124, 126, 128, 129, 137–139, 141–143, 145, 152, 153, 156–158, 162, 187, 190–195, 201, 204, 205, 208, 221, 224, 225
Stoa, Stoics, 88, 108–110, 112, 134, 167, 223
Structure, 2, 3, 7, 19, 20, 60, 61, 64, 66, 77, 85, 87, 90, 97, 100, 107, 108, 115, 117, 120, 121, 128, 135, 137, 139, 143, 145, 151, 157–160, 162, 167–169, 171, 177, 183, 185–187, 198, 222
Subject, 3, 12, 38, 39, 41, 42, 47, 51, 54, 60, 68, 76, 77, 80, 83, 92, 93, 96, 97, 111, 115, 116, 118, 128, 130–133, 135, 137–144, 156, 161–163, 165, 167–170, 180–185, 193, 201, 205, 206, 216, 219, 224, 225
Substance, 59, 119, 133, 153, 154, 200
Synthesis, 2, 79, 129, 143
System, 1, 2, 46, 60, 76, 78, 80, 83, 85, 90, 91, 95, 98, 105, 108, 111, 113–121, 126, 127, 130–132, 135, 137–139, 141, 142, 155, 163, 164, 171, 180–184, 186, 187, 195, 198, 216, 221–223, 226

T

Technē, technical, technology, 28, 54, 134, 136, 216, 230
Teleology, 77–80, 82, 84–86, 88, 145
Temporal modes, 151, 152, 154, 155, 157–160, 171
Theodicy, 220, 221
Theology, 32, 35, 71, 72, 94, 118, 119, 125, 129, 130, 134, 152, 163, 208, 220
Thought, thinking, 19
Transcendence, transcendental, 3, 75–87, 90–105, 126, 130–132, 142, 143, 154, 159, 160, 168, 185, 186, 223, 227
Transcendental semblance, 3, 126, 127, 143, 164, 221, 223
Truth, 1–4, 7–10, 13–22, 24–28, 30–36, 38–42, 44, 45, 48, 49, 52, 53, 56, 60, 67, 72, 94, 97, 116, 118, 119, 123–126, 128, 131, 133–135, 139, 141–147, 151, 157, 159, 165–169, 177–179, 183, 185, 192, 193, 195, 200, 201, 213, 217, 222, 223, 227

U

Unity, 3, 26, 31–33, 36, 40–42, 65, 78, 79, 84, 85, 109, 133, 139, 145, 151–155, 157, 159, 160, 171, 193, 199, 202, 207–209, 225
Universe, universal, 49, 51, 64, 67, 75, 78–85, 87, 88, 90, 91, 93, 95–100, 104, 107–113, 115, 117, 118, 120, 130–133, 141–143, 151, 153, 158–160, 163, 164, 166, 171, 198, 199, 212, 224, 225

V

Value, 45, 48, 51, 53, 96, 101, 117, 135, 145, 146, 181, 182, 195, 216

W

War, 76, 88–90, 99, 104, 125, 138, 140, 204, 207, 216, 226
Will, 96
World, 7, 8, 30, 47, 49, 50, 57, 71, 72, 75–78, 82, 84, 87, 89, 90, 94, 99, 104, 107, 109, 110, 115, 117–120, 123, 125, 126, 131, 135, 136, 140, 142, 145, 152, 153, 155, 163, 166–169, 171, 178, 184, 187, 189, 191, 192, 194, 195, 198–203, 205, 208, 209, 212, 213, 216, 217, 219, 220, 222, 226, 227

Z

Zoē, 166

FSC
www.fsc.org
MIX
Papier | Fördert gute Waldnutzung
FSC® C083411

Kolibri 360 GmbH
Ferdinand-Jühlke-Straße 7
99095 Erfurt, Deutschland
produktsicherheit@kolibri360.de